Getting Started with

MATLAB

Updated for Version 7.8 (Release 2009a)

Getting Started with
MATLAB

A Quick Introduction for
Scientists and Engineers

RUDRA PRATAP

Department of Mechanical Engineering
Indian Institute of Science, Bangalore

New York · Oxford
OXFORD UNIVERSITY PRESS
2010

Oxford University Press, Inc., publishes works that further Oxford University's objective of excellence in research, scholarship, and education.

Oxford   New York
Auckland   Bangkok   Buenos Aires   Cape Town   Chennai
Dar es Salaam   Delhi   Hong Kong   Istanbul   Karachi   Kolkata
Kuala Lumpur   Madrid   Melbourne   Mexico City   Mumbai
Nairobi   São Paulo   Shanghai   Taipei   Tokyo   Toronto

Published by Oxford University Press, Inc.
198 Madison Avenue, New York, New York, 10016
http://www.oup-usa.org

Oxford is a registered trademark of Oxford University Press
MATLAB® is a registered trademark of The MathWorks
Handle Graphics® is a registered trademark of The MathWorks

Library of Congress Cataloging-in-Publication Data

Pratap, Rudra, 1964-
  Getting started with MATLAB : a quick introduction for scientists and engineers / Rudra Pratap.
      p. cm.
ISBN: 978-0-19-973124-4
  1. MATLAB. 2. Science—Data processing. 3. Engineering mathematics—Data processing. I. Title.
  Q183.9.P734 2010
  620.001'51--dc22

                                    2009028033

Printed in the United States of America

10 9 8 7 6 5

To Ma Gayatri

and my parents
Shri Chandrama Singh and Smt. Bachcha Singh

Contents

Preface

I enjoy MATLAB, and I want you to enjoy it too—that is the singular motivation behind this book. The first and foremost goal of this book is to get you started in MATLAB quickly and pleasantly.

Learning MATLAB changed the meaning of scientific computing for me. I used to think in terms of machine-specific compilers and tables of numbers as output. Now, I expect and enjoy interactive calculation, programming, graphics, animation, and complete portability across platforms—all under one roof. MATLAB is simple, powerful, and for most purposes quite fast. This is not to say that MATLAB is free of quirks and annoyances. It is not a complete miracle drug, but I like it and I think you will probably like it too.

I first used MATLAB in 1988 in a course on matrix computation taught by Tom Coleman at Cornell University. We used the original 1984 commercial version of MATLAB. Although the graphics capability was limited to bare-bones 2-D plots, and programming was not possible on the mainframe VAX, I still loved it. Ever since, I have used MATLAB for all my computational needs, for all my work, and in all the courses that I have taught. I have given several introductory lectures, demonstrations, and hands-on workshops. This book is a result of my involvement with MATLAB teaching, both informal and in the classroom, over the last several years.

This book has been around for 15 years now. The fifth edition is in your hand. With every new edition, I face a dilemma—the temptation to add more material and my stubborn desire to keep it lean and thin. I have always tried to strike a balance. This book is not meant to be a manual or an exhaustive account of what MATLAB can do; it is meant to be a friendly introduction that can get you going quickly. Any software package as powerful as MATLAB is likely to have hundreds, if not thousands, of pages of documentation, both on-line and printed. In my experience, what a beginner needs is a filtered set of instructions and discussion that makes learning inviting, fun, and productive. Toward this goal, I have poured my two decades of experience with teaching and MATLAB computation into the pages that follow.

This book is intended to get you started quickly. After an hour or two of getting started, you can use the book as a reference. There are many examples, which you can modify for your own use. The coverage of topics is based on my experience of what is most useful, and what I wish I could have found in a book when I was learning MATLAB. Over the years, I have received numerous feedbacks on this book. Invariably, the chapter on tutorials (Chapter 2) has been hailed as the greatest strength of this book. Therefore, in this edition, I have strengthened that chapter by adding a few more tutorials that invite the reader to explore five different aspects of MATLAB computing. Chapter 2 is now divided into two parts—the basics (five tutorials) and the directional explorations (six tutorials). The basics are meant to get you going within an hour, if you are a first-time user. Then, take a coffee break, and dive into more substantial tutorials of your choice—on arrays, anonymous functions, symbolic mathematics, exporting and importing data, navigating files and directories, or publishing reports. You do not have to go through the later tutorials serially. You can pick and choose. You can also come back to them later when you need to explore that particular aspect of MATLAB computing.

Another major change in this edition is the chapter on Computer Algebra and Symbolic Math Toolbox, Chapter 8. MATLAB has changed the symbolic math engine from Maple to MuPAD. Although, this change is largely unnoticeable to casual users of the Symbolic Math Toolbox, it has prompted me to revise that chapter significantly. I have added some material that brings out the symbolic computation power available to the user by direct access to MuPAD functionality through the MuPAD notebook interface in MATLAB. In particular, I have tried to draw the attention of the reader to the powerful graphics capabilities of MuPAD.

The current edition has been updated for MATLAB Release 2009a. Every update requires checking each command and function given in this book as examples, and changing them if required. One peculiar problem with bringing out a new edition of a book like this (to keep up with the new version of the software package) is to decide which aspects of software upgrade should be included. The new versions of software packages usually add features that their experienced users ask for. As a result, the packages and their manuals get bigger and bigger, and more intimidating to a new user. I have tried hard to protect the interests of a new user in this book. To a new or an average user, most of the distinction in new releases of any software nowadays has to do with look and feel of the software, that is, the user interface. Most of that has to do with rearranged windows, menus, etc.; pretty much like new models of cars–most of the changes are in head-lights, tail-lights, mirrors, etc., not many in engines. Our focus on MATLAB is as a scientific computing and visualization tool. Therefore, I have chosen not to pay much attention to user interface features. I limit the attention to those features that a beginner cannot avoid noticing. If I could, I would like to keep the book largely free of MATLAB screen shots (and thus the dependency on the twice-a-year release of MATLAB versions). Almost every single command or feature discussed in this book should work just fine with MATLAB 2008 and 2009 releases. I do not expect any major changes in most commands and functions presented in this book with the new releases of MATLAB over the next couple of years. However, I do intend to keep a current list of changes on this book's website (**www.oup.com/us/pratap**) to safeguard your interest.

Your feedback is very important to me. If you find the book informative and useful, it is my pleasure to be of service to you. If you find it frustrating, please share your frustrations with me so that I can try to improve future editions.

Acknowledgments

I was helped through the development of this book by the encouragement, criticism, editing, typing, and test-learning of many people, especially at Cornell University and the Indian Institute of Science. I thank all students who used this book in its past forms and provided constructive criticism. I have also been fortunate to receive feedback by email, sometimes quite flattering, from several readers all over the world. I greatly appreciate your words of encouragement.

I wish to thank Chris Wohlever, Mike Coleman, Richard Rand, David Caughey, Yogendra Simha, Vijay Arakeri, Greg Couillard, Christopher D. Hall, James R. Wohlever, John T. Demel, Jeffrey L. Cipolla, John C. Polking, Thomas Vincent, John Gibson, Sai Jagan Mohan, Konda Reddy, Sesha Sai, Yair Hollander, Les Axelrod, Shishir Kumar, The MathWorks Inc., and Cranes Software International Limited for the help and support they have extended to me in the development of this book. In addition, I must acknowledge the help of a few special people. Andy Ruina has been an integral part of the development of this book all along. In fact, he initially wrote most of Chapter 8, the introduction to the Symbolic Math Toolbox. That apart, his criticisms and suggestions have influenced every page of this book. Mohammed Ashraf created script files containing commands and programs from each chapter for checking compatibility with MATLAB 7. Abhay, an energetic student from my lab, has modified those files to make them cell scripts and meticulously checked the integrity and compatibility of all commands and codes given in the book with MATLAB release 2008b and 2009a. He also convinced me to add more material to the symbolic math discussion in Chapter 8, in particular, the introduction to MuPAD's powerful graphics. I thank Manjula for her help with graphics files and proofreading. I also acknowledge the continuous support and encouragement from my editors, Rachael Zimmermann and Patrick Lynch, at the Oxford University Press in bringing out this new edition.

I also thank my wife, Kalpana, and my kids, Manisha, Manas, and Mayank, for being incredibly patient and supportive. The kids have seen their entire summer vacation (of 2009) slip by with their dad spending most of the weekends in front of the computer. Their "are you done yet"s, spoken or otherwise, have made me work much harder. I have seen similar expressions in the eyes of my graduate students. I am thankful to them all for their patience and understanding with my overwhelmingly busy schedule and the consequent unavailability during the last couple of months.

Thank you all,

Bangalore   Rudra Pratap
June 2009

1. *Introduction*

1.1   What Is MATLAB?

MATLAB is a software package for high-performance numerical computation and visualization. It provides an interactive environment with hundreds of built-in functions for technical computation, graphics, and animation. Best of all, it also provides easy extensibility with its own high-level programming language. The name MATLAB stands for MATrix LABoratory.

The diagram in Fig. 1.1 shows the main features and capabilities of MATLAB. MATLAB's built-in functions provide excellent tools for linear algebra computations, data analysis, signal processing, optimization, numerical solution of ordinary differential equations (ODEs), quadrature, and many other types of scientific computations. Most of these functions use state-of-the-art algorithms. There are numerous functions for 2-D and 3-D graphics, as well as for animation. Also, for those who cannot do without their Fortran or C codes, MATLAB even provides an external interface to run those programs from within MATLAB. The user, however, is not limited to the built-in functions; he can write his own functions in the MATLAB language. Once written, these functions behave just like the built-in functions. MATLAB's language is very easy to learn and to use.

There are also several *optional* "toolboxes" available from the developers of MATLAB. These toolboxes are collections of functions written for special applications such as symbolic computation, image processing, statistics, control system design, and neural networks. The list of toolboxes keeps growing with time. There are now more than 50 such toolboxes. We do not attempt introduction to any toolbox here, with the exception of the Symbolic Math Toolbox (Chapter 8).

The basic building block of MATLAB is the matrix. The fundamental data type is the *array*. Vectors, scalars, real matrices, and complex matrices are all automatically handled as special cases of the basic data type. What is more, you almost never have to declare the dimensions of a matrix. MATLAB simply loves matrices

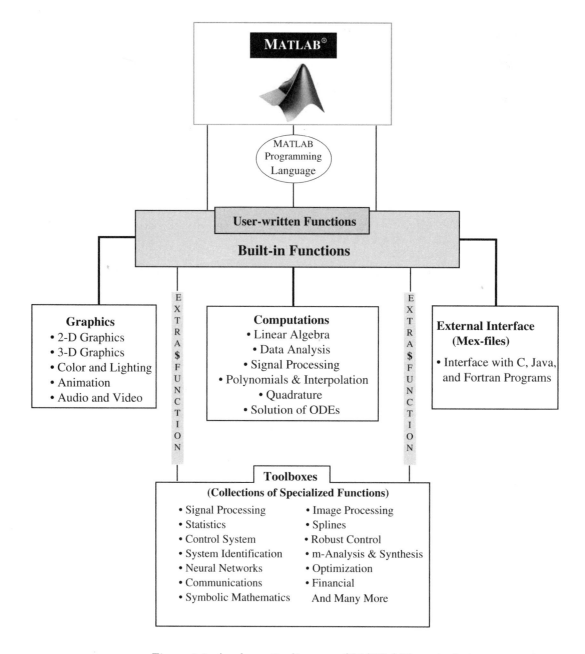

Figure 1.1: A schematic diagram of MATLAB's main features.

and matrix operations. The built-in functions are optimized for vector operations. Consequently, *vectorized*[1] commands or codes run much faster in MATLAB.

1.2   Does MATLAB Do Symbolic Calculations?

(MATLAB vs. Mathematica or Maple)

If you are new to MATLAB, you are likely to ask this question. The first thing to realize is that MATLAB is primarily a numerical computation package, although with the Symbolic Math Toolbox (standard with the Student Edition of MATLAB, see Chapter 8 for an introduction) it can do symbolic algebra.[2] Mathematica and Maple are primarily symbolic algebra packages. Of course, they do *numerical* computations too. In fact, if you know any of these packages *really* well, you can do almost every calculation that MATLAB does using that software. So why learn MATLAB? Well, MATLAB's ease of use is its best feature. Also, it has a shallow learning curve (more learning with less effort) whereas the computer algebra systems have a steep learning curve. Because MATLAB was primarily designed to do numerical calculations and computer algebra systems were not, MATLAB is often much faster at these calculations—often as fast as C or Fortran. There are other packages, such as Xmath, that are also closer in aim and scope but seem to be popular with people in some specialized application areas. The bottom line is, in numerical computations, especially those that use vectors and matrices, MATLAB beats everything hands down in terms of ease of use, availability of built-in functions, ease of programming, and speed. The proof is in the phenomenal growth of MATLAB users around the world in the last two decades. There are more than 2000 universities and thousands of companies listed as registered users. MATLAB's popularity today has forced such powerful packages as Mathematica and many others to provide extensions for files in MATLAB's format!

1.3   Will MATLAB Run on My Computer?

The most likely answer is "yes," because MATLAB supports almost every computational platform. In addition to Windows, MATLAB is available for UNIX, Sun Solaris, Linux, and Mac OS X operating systems. Older versions of MATLAB are available for additional platforms such as Mac OS and Open VMS. To find out more about product availability for your particular computer, see the MathWorks website listed in Section 1.4.

---

[1] *Vectorization* refers to a manner of computation in which an operation is performed simultaneously on a list of numbers (a vector) rather than sequentially on each member of the list. For example, let θ be a list of 100 numbers. Then $y = \sin(\theta)$ is a vectorized statement as opposed to $y_1 = \sin(\theta_1), y_2 = \sin(\theta_2)$, etc.

[2] *Symbolic algebra* means that computation is done in terms of symbols or variables rather than numbers. For example, if you type (x+y)^2 on your computer and the computer responds by saying that the expression is equal to $x^2 + 2xy + y^2$, then your computer does symbolic algebra. Software packages that do symbolic algebra are also known as *computer algebra systems*.

1.4   Where Do I Get MATLAB?

MATLAB is a product of The MathWorks, Incorporated. Contact the company for product information and ordering at the following address:

The MathWorks Inc.
3 Apple Hill Drive, Natick, MA 01760-2098
Phone: (508) 647-7000,   Fax: (508) 647-7001
Email: info@mathworks.com
World Wide Web: http://www.mathworks.com

1.5   How Do I Use This Book?

This book is intended to serve as an introduction to MATLAB. The goal is to get you started as simply as possible. MATLAB is a very powerful and sophisticated package. It takes a while to understand its real power. Unfortunately, most powerful packages tend to be somewhat intimidating to a beginner. That is why this book exists—to help you overcome the fear, get started quickly, and become productive in very little time. The most useful and easily *accessible* features of MATLAB are discussed first to make you productive and build your confidence. Several features are discussed in sufficient depth, with an invitation to explore the more advanced features on your own. All features are discussed through examples using the following conventions:

- **Typographical styles:**

    - All actual MATLAB commands or instructions are shown in `typed face`. Menu commands, files names, etc., are shown in sans serif font.
    - Place holders for variables or names in a command are shown in *italics*. So, a command shown as `help` *`topic`* implies that you have to type the actual name of a topic in place of *topic* in the command.
    - *Italic* text has also been used to *emphasize* a point and, sometimes, to introduce a new term.

- **Actual examples:** Actual examples carried out in MATLAB are shown in gray, shaded boxes. Explanatory notes have been added within small white rectangles in the gray boxes, as shown in Fig. 1.2. These gray, boxed figures are intended to provide a parallel track for the impatient reader. If you would rather try out MATLAB right away, you are encouraged to go through these boxed examples. Most of the examples are designed so that you can (more or less) follow them without reading the entire text. All examples are system-independent. After trying out the examples, you should read the appropriate sections.

For on-line help type:
`help` *`topic`*

- **On-line help:** We encourage the use of on-line help. For almost all major topics, we indicate the on-line help information in a small box in the margin, as shown here on the left.

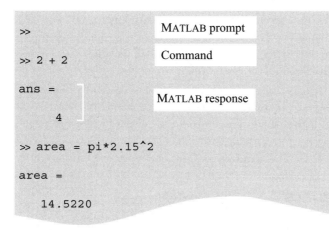

Figure 1.2: Actual examples carried out in MATLAB are shown in gray boxes throughout this book. The text in the white boxes inside these gray boxes are explanatory notes.

Typing `help` *topic* in MATLAB with the appropriate topic name provides a list of functions and commands for that topic. Detailed help can then be obtained for any of those commands and functions.

We discourage a passive reading of this book. The best way to learn any computer software is to try it out. We believe this, practice it, and encourage you to practice it too. So, if you are impatient, quickly read Sections 1.6.1–1.6.3, jump to the tutorials on page 15, and get going.

1.6 Basics of MATLAB

Here we discuss some basic features and commands. To begin, let us look at the general structure of the MATLAB environment.

1.6.1 MATLAB windows

On almost all systems, MATLAB works through three basic windows, which are shown in Fig. 1.3 and discussed here.

1. **MATLAB desktop:** This is where MATLAB puts you when you launch it (see Fig. 1.3). The MATLAB desktop, by default, consists of the following subwindows.

   Command window: This is the main window. It is characterized by the MATLAB command prompt ($\gg$). When you launch the application program, MATLAB puts you in this window. All commands, including those for running user-written programs, are typed in this window at the

MATLAB prompt. In MATLAB, this window is a part of the MATLAB window (see Fig. 1.3) that contains other smaller windows or *panes*. If you can get to the command window, we advise you to ignore the other four subwindows at this point. As software packages become more and more powerful, their creators add more and more features to address the needs of experienced users. Unfortunately, it makes life harder for the beginners—there is more room for confusion, distraction, and intimidation. Although we describe the other subwindows here that appear with the command window, we do not expect it to be useful to you till you get to Lesson 3 in Chapter 2.

Current Directory pane: This pane is located on the left of the **Command Window** in the default MATLAB desktop layout. This is where all your files from the current directory are listed. You can do file navigation here. Make sure that this is the directory where you want to work so that MATLAB has access to your files and where it can save your new files. If you change the current directory (by navigating through your file system), make sure that the selected directory is also reflected in the little window above the **Command Window** marked **Current Directory**. This little window and the current directory pane are interlinked; changing the directory in one is automatically reflected in the other.

You also have several options of what you can do with a file once you select it (with a mouse click). To see the options, click the right button of the mouse after selecting a file. You can run M-files, rename them, delete them, etc.

(File) Details pane: Just below the **Current Directory** pane is the **Details** pane that shows the details of a file you select in the current directory pane. These details are normally limited to listing of variables from a MAT-file (a binary data file discussed later), showing titles of M-files, and listing heading of cells if present in M-files. You do not need to understand these details yet.

Workspace pane: This subwindow lists all variables that you have generated so far and shows their type and size. You can do various things with these variables, such as plotting, by clicking on a variable and then using the right button on the mouse to select your options.

Command History pane: All commands typed on the MATLAB prompt in the command window get recorded, even across multiple sessions (you worked on Monday, then on Thursday, and then on next Wednesday, and so on), in this window. You can select a command from this window with the mouse and execute it in the command window by double-clicking on it. You can also select a set of commands from this window and create an M-file with the right click of the mouse (and selecting the appropriate option from the menu).

2. **Figure window:** The output of all graphics commands typed in the command window are flushed to the graphics or figure window, a separate gray window with (default) white background color. The user can create as many figure windows as the system memory will allow.

3. **Editor window:** This is where you write, edit, create, and save your own programs in files called *M-files*. You can use any text editor to carry out these tasks. On most systems, MATLAB provides its own built-in editor. However, you can use your own editor by typing the standard file-editing command that you normally use on your system. From within MATLAB, the command is typed at the MATLAB prompt following the exclamation character (!). The exclamation character prompts MATLAB to return the control temporarily to the local operating system, which executes the command following the character. After the editing is completed, the control is returned to MATLAB. For example, on UNIX systems, typing `!vi myprogram.m` at the MATLAB prompt (and hitting the return key at the end) invokes the *vi* editor on the file `myprogram.m`. Typing `!emacs myprogram.m` invokes the *emacs* editor.

1.6.2 On-line help

- **On-line documentation:** MATLAB provides on-line help for all its built-in functions and programming language constructs. The commands `lookfor`, `help`, `helpwin`, and `helpdesk` provide on-line help. See Section 3.6 on page 85 for a description of the help facility.

- **Demo:** MATLAB has a demonstration program that shows many of its features. The program includes a tutorial introduction that is worth trying. Type `demo` at the MATLAB prompt to invoke the demonstration program, and follow the instructions on the screen.

1.6.3 Input-output

MATLAB supports interactive computation (see Chapter 3), taking the input from the screen, and flushing the output to the screen. In addition, it can read input files and write output files (see Section 4.3.7). The following features hold for all forms of input-output:

- **Data type:** The fundamental data type in MATLAB is an *array*. It encompasses several distinct data objects—integers, doubles (real numbers), matrices, character strings, structures, and cells.[3] In most cases, however, you never have to worry about the data type or the data object declarations. For example, there is no need to declare variables as real or complex. When a real number is entered as the value of a variable, MATLAB automatically sets the variable to be real (`double`).

---

[3] *Structures* and *cells* as data objects were introduced in MATLAB 5. See Section 4.4 on page 121 for their description. MATLAB also allows users to create their own data objects and associated operations. We do not discuss this facility in this book.

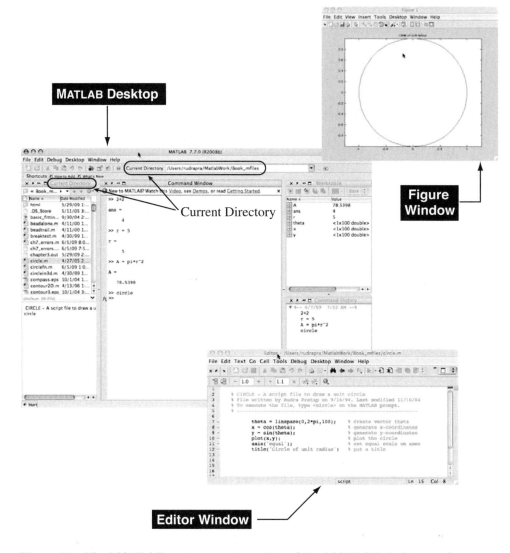

Figure 1.3: The MATLAB environment consists of the MATLAB desktop, a figure window, and an editor window. The figure and the editor windows appear only when invoked with the appropriate commands. For example, you can open the editor window by selecting File → New → Blank M-File, and open a blank figure window by selecting File → New → Figure.

- **Dimensioning:** Dimensioning is automatic in MATLAB. No dimension statements are required for vectors or arrays. You can find the dimensions of an existing matrix or a vector with the `size` and `length` (for vectors only) commands.

- **Case sensitivity:** MATLAB is case-sensitive; that is, it differentiates between the lowercase and uppercase letters. Thus `a` and `A` are different variables. Most MATLAB commands and built-in function calls are typed in lowercase letters. You can turn case sensitivity on and off with the `casesen` command. However, we do not recommend it.

- **Output display:** The output of every command is displayed on the screen unless MATLAB is directed otherwise. A semicolon at the end of a command suppresses the screen output, except for graphics and on-line help commands. The following facilities are provided for controlling the screen output:

    - **Paged output:** To direct MATLAB to show one screen of output at a time, type `more on` at the MATLAB prompt. Without it, MATLAB flushes the entire output at once, without regard to the speed at which you read.

    - **Output format:**

      For on-line help type: `help format`

      Though computations inside MATLAB are performed using double precision, the appearance of floating point numbers on the screen is controlled by the output format in use. There are several different screen output formats. The following table shows the printed value of 10π in seven different formats.

      ```
      format short       31.4159
      format short e     3.1416e+001
      format long        31.41592653589793
      format long e      3.141592653589793e+001
      format short g     31.416
      format long g      31.4159265358979
      format hex         403f6a7a2955385e
      format rat         3550/113
      format bank        31.42
      ```

      The additional formats, `format compact` and `format loose`, control the spacing above and below the displayed lines, and `format +` displays a `+`, `-`, and blank for positive, negative, and zero numbers, respectively. The default is `format short`. The display format is set by typing `format` *type* on the command line (see Fig. 2.1 on page 18 for an example).

- **Command history:** MATLAB saves previously typed commands in a buffer. These commands can be recalled with the up-arrow key (↑). This helps in editing previous commands. You can also recall a previous command by typing the first few characters and then pressing the ↑ key. Alternatively, you can double-click on a command in the **Command History** pane (where all your commands from even previous sessions of MATLAB are recorded and listed) to execute it in the command window. On most UNIX systems, MATLAB's command-line editor also understands the standard *emacs* keybindings.

1.6.4   File types

For on-line help type:
`help fileformats`

MATLAB can read and write several types of files. However, there are mainly five different types of files for storing data or programs that you are likely to use often:

M-files are standard ASCII text files, with a .m extension to the filename. There are two types of these files: *script files* and *function files* (see Sections 4.1 and 4.2). Most programs you write in MATLAB are saved as M-files. All built-in functions in MATLAB are M-files, most of which reside on your computer in precompiled format. Some built-in functions are provided with source code in readable M-files so that they can be copied and modified.

Mat-files are binary datafiles, with a .mat extension to the filename. Mat-files are created by MATLAB when you save data with the `save` command. The data is written in a special format that only MATLAB can read. Mat-files can be loaded into MATLAB with the `load` command (see Section 3.7 for details).

Fig-files are binary figure files with a .fig extension that can be opened again in MATLAB as figures. Such files are created by saving a figure in this format using the **Save** or **Save As** options from the **File** menu or using the `saveas` command in the command window. A fig-file contains all the information required to recreate the figure. Such files can be opened with the `open filename.fig` command.

P-files are compiled M-files with a .p extension that can be executed in MATLAB directly (without being parsed and compiled). These files are created with the `pcode` command. If you develop an application that other people can use but you do not want to give them the source code (M-file), then you give them the corresponding p-code or the p-file.

Mex-files are MATLAB-callable Fortran, C, and Java programs, with a .mex extension to the filename. Use of these files requires some experience with MATLAB and a lot of patience. We do not discuss Mex-files in this introductory book.

1.6.5   Platform dependence

One of the best features of MATLAB is its platform independence. Once you are in MATLAB, for the most part, it does not matter which computer you are on. Almost all commands work the same way. The only commands that differ are the ones that necessarily depend on the local operating system, such as editing (if you do not use the built-in editor) and saving M-files. Programs written in the MATLAB language work exactly the same way on all computers. The user interface (how you interact with your computer), however, may vary a little from platform to platform.

- **Launching MATLAB:** If MATLAB is installed on your machine correctly then you can launch it by following these directions:

  On PCs: Navigate and find the MATLAB folder, locate the MATLAB program, and double-click on the program icon to launch MATLAB. If you have worked in MATLAB before and have an M-file or Mat-file that was written by MATLAB, you can also double-click on the file to launch MATLAB.

  On UNIX machines: Type `matlab` on the UNIX prompt and hit return or enter. If MATLAB is somewhere in your *path*, it will be launched. If it is not, ask your system administrator.

- **Creating a directory and saving files:** Where should you save your files so that MATLAB can easily access them? MATLAB creates a default folder called **Matlab** inside **Documents** (on Macs), or **My Documents** (on PCs) where it saves your files if you do not specify any other location. If you are the only user of MATLAB on the computer you are working on, this is fine. You can save all your work in this folder and access all your files easily (default setup). If not, you have to create a separate folder for saving your work.

  Theoretically, you can create a directory/folder anywhere, save your files, and direct MATLAB to find those files. The most convenient place, however, to save all user-written files is in the default directory **MATLAB** created by the application in your **Documents** or **My Documents** folder. This way all user-written files are automatically accessible to MATLAB. If you need to store the files somewhere else, you might have to specify the path to the files using the `path` command, or change the working directory of MATLAB to the desired directory with a few navigational clicks in the **Current Directory** pane. We recommend the latter.

- **Printing:**

  On PCs: To print the contents of the current active window (command, figure, or edit window), select **Print...** from the **File** menu and click **Print** in the dialog box. You can also print the contents of the figure window by typing `print` at the MATLAB prompt.

  On UNIX machines: To print a file from inside MATLAB, type the appropriate UNIX command preceded by the exclamation character (!). For example, to print the file **startup.m**, type `!lpr startup.m` on the MATLAB prompt. To print a graph that is currently in the figure window simply type `print` on the MATLAB prompt.

1.6.6   General commands you should remember

On-line help

help	lists topics on which help is available
helpwin	opens the interactive help window
helpdesk	opens the web browser-based help facility
help *topic*	provides help on *topic*
lookfor *string*	lists help topics containing *string*
demo	runs the demo program

Workspace information

who	lists variables currently in the workspace
whos	lists variables currently in the workspace with their size
what	lists M-, Mat-, and Mex-files on the disk
clear	clears the workspace, all variables are removed
clear x y z	clears only variables x, y, and z
clear all	clears all variables and functions from workspace
mlock *fun*	locks function *fun* so that **clear** cannot remove it
munlock *fun*	unlocks function *fun* so that **clear** can remove it
clc	clears command window, cursor moves to the top
home	scrolls the command window to put the curser on top
clf	clears figure window

Directory information

pwd	shows the current working directory
cd	changes the current working directory
dir	lists contents of the current directory
ls	lists contents of the current directory, same as **dir**
path	gets or sets MATLAB search path
editpath	modifies MATLAB search path
copyfile	copies a file
mkdir	creates a directory

General information

computer	tells you the computer type you are using
clock	gives you wall clock time and date as a vector
date	tells you the date as a string
more	controls the paged output according to the screen size
ver	gives the license and the MATLAB version information
bench	benchmarks your computer on running MATLAB compared to other computers

Termination

^c (Control-c)	local abort, kills the current command execution
quit	quits MATLAB
exit	same as **quit**

1.7   Visit This Again

We would like to point out a few things that vex the MATLAB beginners, perhaps, the most. Although many of these things would probably not make sense to you right now, they are here, and you can come back to them whenever they seem relevant.

In the past, file navigation in MATLAB has caused considerable problems for users, especially the beginners. We have had numerous complaints from students about not being able to make MATLAB find their file, get MATLAB to work from their directory, get MATLAB to find and execute the currently edited file, etc. Fortunately, from MATLAB 6 onward, MathWorks has incorporated several new features that mitigate this problem immensely. The current directory is shown just above the command window with the option of changing it with just a click of the mouse. There is also a current directory subwindow to the left of the command window that lists files in the current directory and gives you options of opening, loading (Mat-file), executing (M-file), editing, etc., with the click of the right button on the mouse. You can also change the directory there or add a particular directory to the MATLAB path so that MATLAB has access to all the files in that directory automatically.

If you do not save all your MATLAB files in the default **Work** directory or folder, you need to be aware of the following issues.

1. **Not being in the right directory:** You may write and save many MATLAB programs (M-files) but MATLAB does not seem to find them. If your files are not in the current working directory, MATLAB cannot access them. Find which directory you are currently in by looking at the small current directory window in the toolbar or by querying MATLAB with the command `pwd`. If you are not in the right place, guide MATLAB to get to the directory where your files are. See Lesson 10 in the tutorials (Chapter 2).

2. **Not saving files in the correct directory:** When you edit a file in the MATLAB editor/debugger window and save it, it does not automatically mean that the MATLAB command window has access to the directory you saved your file in. So, after saving the file, when you try to execute it and MATLAB does not find your file, follow item 1 above and set things right.

3. **Not overwriting an existing file while editing:** You run your program by executing your M-file, do not like the result, edit the file, and run it again; but MATLAB gives the same answer! The previously *parsed* (compiled) file is executing; MATLAB does not know about your changes. This can happen due to various reasons. The simple cure is, clear the workspace with `clear all` and execute your file.

There are various other little things that cause trouble from time to time. We point them out throughout the book wherever they raise issues.

$\mathcal{2.}$ Tutorial Lessons

The following lessons are designed to get you started quickly in MATLAB. Each lesson should take about 10–15 minutes. The lessons are intended to make you familiar with the basic facilities of MATLAB. We also urge you to do the exercises given at the end of each lesson. This will take more time, but it will make you familiar with MATLAB. If you get stuck in the exercises, simply turn the page; answers are on the back. Most answers consist of correct commands to do the exercises. However, there are several correct ways to do the problems. So, your commands may look different than those given.

Before You Start

You need some information about the computer you are going to work on. In particular, find out the following:

- How to switch on the computer and get it started.
- How to log on and log off.
- Where MATLAB is installed on the computer.
- How to access MATLAB.
- Where you can write and save files—hard drive or a floppy disk.
- If there is a printer attached to the computer.

If you are working on your own computer, you will most likely know the answer to these questions. If you are working on a computer in a public facility, the system manager can help you. If you are in a class that requires working on MATLAB, your professor or TA can provide answers. In public facilities, sometimes the best thing to do is to spot a friendly person who works there and ask these questions politely. People are usually nice!

If you have not read the introduction (Chapter 1), we recommend that you at least read Sections 1.6.1–1.6.3 and glance through the rest of Section 1.6 before trying the tutorials.

The Basics

Here are the lessons in a nutshell:

Lesson 1: Launch MATLAB, do some simple calculations, and quit.

Key features: Learn to add, multiply, and exponentiate numbers; use trigonometric functions; and control screen output with `format`.

Lesson 2: Create and work with arrays, vectors in particular.

Key features: Learn to create, add, and multiply vectors; use `sin` and `sqrt` functions with vector arguments; and use `linspace` to create a vector.

Lesson 3: Plot simple graphs.

Key features: Learn to plot, label, and print out a circle.

Lesson 4: Write and execute a *script file*.

Key features: Learn to write, save, and execute a script file that plots a unit circle.

Lesson 5: Write and execute a *function file*.

Key features: Learn to write, save, and execute a function file that plots a circle of any specified radius.

2.1   Lesson 1: A Minimum MATLAB Session

Goal: To learn how to log on, invoke MATLAB, do a few trivial calculations, quit MATLAB, and log off.

Time Estimates

Lesson: 10 minutes

Exercises: 30 minutes

What you are going to learn

- How to do simple arithmetic calculations. The arithmetic operators are

+	addition,
−	subtraction,
*	multiplication,
/	division, and
^	exponentiation.

- How to assign values to variables.
- How to suppress screen output.
- How to control the appearance of floating point numbers on the screen.
- How to quit MATLAB.

The MATLAB commands/operators used are

```
+, -, *, /, ^, ;
sin, cos, log
format
quit
```

In addition, if you do the exercises, you will learn more about arithmetic operations, exponentiation and logarithms, trigonometric functions, and complex numbers.

Method: Log on and launch MATLAB. Once the MATLAB command window is on the screen, you are ready to carry out the first lesson. Some commands and their output are shown in Fig. 2.1. Go ahead and reproduce the results.

```
>> 2 + 2

ans =

        4
```

Enter 2+2 and hit the return/enter key. Note that the result of an un-assigned expression is saved in the default variable *ans*.

```
>> x = 2 + 2

x =


        4
```

You can also assign the value of an expression to a variable.

```
>> y = 2^2 + log(pi)*sin(x);

>> y

y =

      3.1337
```

A semicolon at the end suppresses screen output. MATLAB remembers *y*, though. You can recall the value *y* by simply typing *y*.

```
>> theta = acos(-1)

theta =

      3.1416
```

MATLAB knows trigonometry. Here is arccosine of –1.

```
>> format short e
>> theta

theta =

    3.1416e+000

>> format long
>> theta

theta =

    3.141592653589793
```

The floating point output display is controlled by the format command. Here are two examples. More information will be provided on this later.

```
>> quit
```

Quit MATLAB. You can also quit by selecting **Quit** from the file menu on Macs and PCs.

Figure 2.1: Lesson 1: Some simple calculations in MATLAB.

EXERCISES

1. **Arithmetic operations:** Compute the following quantities:
   - $\frac{2^5}{2^5-1}$ and compare with $(1-\frac{1}{2^5})^{-1}$.
   - $3\frac{\sqrt{5}-1}{(\sqrt{5}+1)^2} - 1$. The square root $\sqrt{x}$ can be calculated with the command `sqrt(x)` or `x^0.5`.
   - Area $= \pi r^2$ with $r = \pi^{\frac{1}{3}} - 1$. ($\pi$ is `pi` in MATLAB.)

2. **Exponential and logarithms:** The mathematical quantities e^x, $\ln x$, and $\log x$ are calculated with `exp(x)`, `log(x)`, and `log10(x)`, respectively. Calculate the following quantities:
   - e^3, $\ln(e^3)$, $\log_{10}(e^3)$, and $\log_{10}(10^5)$.
   - $e^{\pi\sqrt{163}}$.
   - Solve $3^x = 17$ for x and check the result. (The solution is $x = \frac{\ln 17}{\ln 3}$. You can verify the result by direct substitution.)

3. **Trigonometry:** The basic MATLAB trigonometric functions are `sin, cos, tan, cot, sec,` and `csc`. The inverses, e.g., arcsin, arctan, etc., are calculated with `asin, atan,` etc. The same is true for hyperbolic functions. The inverse function `atan2` takes two arguments, y and x, and gives the four-quadrant inverse tangent. The argument of these functions must be in radians. Calculate the following quantities:
   - $\sin\frac{\pi}{6}$, $\cos\pi$, and $\tan\frac{\pi}{2}$.
   - $\sin^2\frac{\pi}{6} + \cos^2\frac{\pi}{6}$. (Typing `sin^2(x)` for $\sin^2 x$ will produce an error).
   - $y = \cosh^2 x - \sinh^2 x$, with $x = 32\pi$.

4. **Complex numbers:** MATLAB recognizes the letters i and j as the imaginary number $\sqrt{-1}$. A complex number $2 + 5i$ may be input as `2+5i` or `2+5*i` in MATLAB. The former case is always interpreted as a complex number, whereas the latter case is taken as complex only if i has not been assigned any local value. The same is true for j. This kind of context dependence, for better or worse, pervades MATLAB. Compute the following quantities:
   - $\frac{1+3i}{1-3i}$. Can you check the result by hand calculation?
   - $e^{i\frac{\pi}{4}}$. Check the Euler's Formula $e^{ix} = \cos x + i\sin x$ by computing the right-hand side too, i.e., compute $\cos(\pi/4) + i\sin(\pi/4)$.
   - Execute the commands `exp(pi/2*i)` and `exp(pi/2i)`. Can you explain the difference between the two results?

Answers to Exercises

1. **Command**   **Result**
   ```
   2^5/(2^5-1)
   ```                                                            1.0323
   ```
   3*(sqrt(5)-1)/(sqrt(5)+1)^2 - 1
   ```                                                            $-0.6459$
   ```
   area=pi*(pi^(1/3)-1)^2
   ```                                                            0.6781

2. **Command**                                                    **Result**
   ```
   exp(3)
   ```                                                            20.0855
   ```
   log(exp(3))
   ```                                                            3
   ```
   log10(exp(3))
   ```                                                            1.3029
   ```
   log10(10^5)
   ```                                                            5
   ```
   exp(pi*sqrt(163))
   ```                                                            2.6254e+17
   ```
   x=log(17)/log(3)
   ```                                                            2.5789

3. **Command**                                                    **Result**
   ```
   sin(pi/6)
   ```                                                            0.5000
   ```
   cos(pi)
   ```                                                            $-1$
   ```
   tan(pi/2)
   ```                                                            1.6331e+16
   ```
   (sin(pi/6))^2+(cos(pi/6))^2
   ```                                                            1
   ```
   x=32*pi; y=(cosh(x))^2-(sinh(x))^2
   ```                                                            0

4. **Command**                                                    **Result**
   ```
   (1+3i)/(1-3i)
   ```                                                            $-0.8000 + 0.6000i$
   ```
   exp(i*pi/4)
   ```                                                            $0.7071 + 0.7071i$
   ```
   exp(pi/2*i)
   ```                                                            $0.0000 + 1.0000i$
   ```
   exp(pi/2i)
   ```                                                            $0.0000 - 1.0000i$

   Note that
   $$\texttt{exp(pi/2*i)} = e^{\frac{\pi}{2}i} = \cos\left(\frac{\pi}{2}\right) + i\,\sin\left(\frac{\pi}{2}\right) = i$$
   $$\texttt{exp(pi/2i)} = e^{\frac{\pi}{2i}} = e^{-\frac{\pi}{2}i} = \cos\left(\frac{\pi}{2}\right) - i\,\sin\left(\frac{\pi}{2}\right) = -i$$

## 2.2 Lesson 2: Creating and Working with Arrays of Numbers

*Goal:* To learn how to create arrays and vectors and how to perform arithmetic and trigonometric operations on them.

An *array* is a list of numbers or expressions arranged in horizontal rows and vertical columns. When an array has only one row or column, it is called a *vector*. An array with $m$ rows and $n$ columns is called a *matrix* of size $m \times n$. See Section 3.1 for more information.

**Time Estimates**

*Lesson:* 15 minutes

*Exercises:* 45 minutes

**What you are going to learn**

- How to create row and column vectors.

- How to create a vector of $n$ numbers linearly (equally) spaced between two given numbers $a$ and $b$.

- How to do simple arithmetic operations on vectors.

- How to do *array operations*:

| | |
|---|---|
| .* | term-by-term multiplication, |
| ./ | term-by-term division, and |
| .^ | term-by-term exponentiation. |

- How to use trigonometric functions with array arguments.

- How to use elementary math functions such as square root, exponentials, and logarithms with array arguments.

This lesson deals primarily with 1-D arrays, i.e., vectors. One of the exercises introduces you to 2-D arrays, i.e., matrices. There are many mathematical concepts associated with vectors and matrices that we do not mention here. If you have some background in linear algebra, you will find that MATLAB is set up to do almost any matrix computation (e.g., inverse, determinant, rank).

*Method:* You already know how to launch MATLAB. Go ahead and try the commands shown in Fig. 2.2. Once again, you are going to reproduce the results shown.

```
>> x = [1 2 3]                          x is a row vector with three elements.

x =
       1       2       3

>> y = [2; 1; 5]
                                        y is a column vector with three
y =                                     elements.
       2
       1
       5

>> z = [2 1 0];                         You can add (or subtract) two
>> a = x + z                            vectors of the same size.

a =
       3       3       3

>> b = x + y                            But you cannot add (or subtract) a
                                        row vector to a column vector.
??? Error using ==> plus
Matrix dimensions must agree.

>> a = x.*z                             You can multiply (or divide) the
                                        elements of two same-sized vectors
a =                                     term by term with the array operator
       2       2       0                .* (or ./) .

>> b = 2*a
                                        But multiplying a vector with a
b =                                     scalar does not need any special
       4       4       0                operation (no dot before the *).

>> x = linspace(0,10,5)                 Create a vector x with 5 elements
                                        linearly spaced between 0 and 10.
x =
          0    2.5000    5.0000    7.5000   10.0000

>> y = sin(x);                          Trigonometric functions sin, cos,
                                        etc., as well as elementary math
>> z = sqrt(x).*y                       functions sqrt, exp, log, etc.,
                                        operate on vectors term by term.
z =
          0    0.9463   -2.1442    2.5688   -1.7203
```

Figure 2.2: Lesson 2: Some simple calculations with vectors.

EXERCISES

1. **Equation of a straight line:** The equation of a straight line is $y = mx + c$, where m and c are constants. Compute the y-coordinates of a line with slope $m = 0.5$ and the intercept $c = -2$ at the following x-coordinates:

$$x = 0, \quad 1.5, \quad 3, \quad 4, \quad 5, \quad 7, \quad 9, \text{ and } 10.$$

   [Note: Your command should not involve any array operators because your calculation involves multiplication of a vector with a scalar m and then addition of another scalar c.]

2. **Multiply, divide, and exponentiate vectors:** Create a vector t with 10 elements: 1, 2, 3, ..., 10. Now compute the following quantities:

   - $x = t \sin(t)$.
   - $y = \frac{t-1}{t+1}$.
   - $z = \frac{\sin(t^2)}{t^2}$.

3. **Points on a circle:** All points with coordinates $x = r \cos\theta$ and $y = r \sin\theta$, where r is a constant, lie on a circle with radius r, i.e., they satisfy the equation $x^2 + y^2 = r^2$. Create a column vector for θ with the values 0, $\pi/4$, $\pi/2$, $3\pi/4$, π, and $5\pi/4$. Take $r = 2$ and compute the column vectors x and y. Now check that x and y indeed satisfy the equation of a circle, by computing the radius $r = \sqrt{(x^2 + y^2)}$. [To calculate r you will need the array operator `.^` for squaring x and y. Of course, you could compute x^2 by `x.*x` also.]

4. **The geometric series:** This is funky! You know how to compute x^n element by element for a vector x and a scalar exponent n. How about computing n^x, and what does it mean? The result, again, is a vector with elements n^{x_1}, n^{x_2}, n^{x_3}, etc. The sum of a geometric series $1 + r + r^2 + r^3 + \cdots + r^n$ approaches the limit $\frac{1}{1-r}$ for $r < 1$ as $n \to \infty$. Create a vector n of 11 elements from 0 to 10. Take $r = 0.5$ and create another vector $x = [r^0 \quad r^1 \quad r^2 \quad \cdots \quad r^n]$ with the `x=r.^n` command. Now take the sum of this vector with the command `s=sum(x)` (s is the sum of the actual series). Calculate the limit $\frac{1}{1-r}$ and compare the computed sum s. Repeat the procedure taking n from 0 to 50 and then from 0 to 100.

5. **Matrices and vectors:** Go to Fig. 3.1 on page 67 and reproduce the results. Now create a vector and a matrix with the following commands: `v=0:0.2:12;` and `M=[sin(v); cos(v)];` (see Section 3.1.4 on page 72 for use of ":" in creating vectors). Find the sizes of v and M using the `size` command. Extract the first 10 elements of each row of the matrix and display them as column vectors.

Answers to Exercises

Commands to solve each problem are given here.

1. ```
   x=[0 1.5 3 4 5 7 9 10];
   y=0.5*x-2
   ```
   [Ans. $y = -2.0000 \quad -1.2500 \quad -0.5000 \quad 0 \quad 0.5000 \quad 1.5000 \quad 2.5000 \quad 3.0000$]

2. ```
   t=1:10;
   x=t.*sin(t)
   y=(t-1)./(t+1)
   z=sin(t.^2)./(t.^2)
   ```

3. ```
   theta=[0;pi/4;pi/2;3*pi/4;pi;5*pi/4]
   r=2;
   x=r*cos(theta);
   y=r*sin(theta);
   x.^2+y.^2
   ```

4. ```
   n=0:10;
   r=0.5;
   x=r.^n;
   s1=sum(x)
   n=0:50;
   x=r.^n;
   s2=sum(x)
   n=0:100;
   x=r.^n;
   s3=sum(x)
   ```
   [Ans. $s1 = 1.9990$, $s2 = 2.0000$, and $s3 = 2$]

5. ```
   v=0:0.2:12;
   M=[sin(v); cos(v)];
   size(v)
   size(M)
   M(:,1:10)'
   ```
   [Ans. v is 1×61 and M is 2×61. The M(:,1:10)' command picks out the first 10 elements from each row of M and transposes them to give a 10×2 matrix.]
   ```

2.3   Lesson 3: Creating and Printing Simple Plots

Goal: To learn how to make a simple 2-D plot in MATLAB and print it out.

Time Estimates
Lesson: 10 minutes
Exercises: 40 minutes

What you are going to learn

- How to generate x- and y-coordinates of 100 equidistant points on a unit circle.

- How to plot x vs. y and thus create the circle.

- How to set the scale of the x-axis and the y-axis to be the same, so that the circle looks like a circle and not an ellipse.

- How to label the axes with text strings.

- How to title the graph with a text string.

- How to get a hard copy of the graph.

The MATLAB commands used are

plot	creates a 2-D line plot,
axis	changes the aspect ratio of the x-axis and the y-axis,
xlabel	annotates the x-axis,
ylabel	annotates the y-axis,
title	puts a title on the plot, and
print	prints a hard copy of the plot.

This lesson teaches you the most basic graphics commands. The exercises take you through various types of plots, overlay plots, and more involved graphics.

Method: You are going to draw a circle of unit radius. To do this, first generate the data (x- and y-coordinates of, say, 100 points on the circle), then plot the data, and finally print the graph. For generating data, use the parametric equation of a unit circle:

$$x = \cos\theta, \quad y = \sin\theta, \quad 0 \le \theta \le 2\pi.$$

In the sample session shown in Fig. 2.3, only the commands are listed. You should see the output on your screen.

```
>> theta = linspace(0,2*pi,100);    Create a linearly spaced 100-
                                     elements-long vector θ.
>> x = cos(theta);
                                     Calculate x- and y-coordinates.
>> y = sin(theta);

>> plot(x,y)                         Plot x vs. y (see Section 6.1).

>> axis('equal');                    Set the length scales of the two
                                     axes to be the same.
>> xlabel('x')                       Label the x-axis with x.

>> ylabel('y')                       Label the y-axis with y.

>> title('Circle of unit radius')   Put a title on the plot.

>> print                             Print on the default printer.
```

Figure 2.3: Lesson 3: Plotting and printing a simple graph.

Comments:

- After you enter the command plot(x,y), you should see an ellipse in the figure window. MATLAB draws an ellipse rather than a circle because of its default rectangular axes. The command axis('equal') directs MATLAB to use the same scale on both axes, so that a circle appears as a circle. You can also use axis('square') to override the default rectangular axes.

- The arguments of the axis, xlabel, ylabel, and title commands are text strings. Text strings are entered within single right-quote (') characters. For more information on text strings, see Section 3.3 on page 77.

- The print command sends the current plot to the printer connected to your computer.

EXERCISES

1. **A simple sine plot:** Plot $y = \sin x$, $0 \leq x \leq 2\pi$, taking 100 linearly spaced points in the given interval. Label the axes and put "Plot created by *yourname*" in the title.

2. **Line styles:** Make the same plot as in Exercise 1, but rather than displaying the graph as a curve, show the unconnected data points. To display the data points with small circles, use `plot(x,y,'o')`. [Hint: You may look into Section 6.1 on page 175 if you wish.] Now combine the two plots with the command `plot(x,y,x,y,'o')` to show the line through the data points as well as the distinct data points.

3. **An exponentially decaying sine plot:** Plot $y = e^{-0.4x} \sin x$, $0 \leq x \leq 4\pi$, taking 10, 50, and 100 points in the interval. [Be careful about computing y. You need array multiplication between `exp(-0.4*x)` and `sin(x)`. See Section 3.2.1 on page 73 for more discussion on array operations.]

4. **Space curve:** Use the command `plot3(x,y,z)` to plot the circular helix $x(t) = \sin t$, $y(t) = \cos t$, $z(t) = t$, $0 \leq t \leq 20$.

5. **On-line help:** Type `help plot` on the MATLAB prompt and hit return. If too much text flashes by the screen, type `more on`, hit return, and then type `help plot` again. This should give you paged screen output. Read through the on-line help. To move to the next page of the screen output, simply press the spacebar.

6. **Log-scale plots:** The plot commands `semilogx`, `semilogy`, and `loglog` plot the x-values, the y-values, and both x- and y-values on a $\log_{10}$ scale, respectively. Create a vector `x=0:10:1000`. Plot x vs. x^3 using the three log-scale plot commands. [Hint: First, compute `y=x.^3` and then use `semilogx(x,y)`, etc.]

7. **Overlay plots:** Plot $y = \cos x$ and $z = 1 - \frac{x^2}{2} + \frac{x^4}{24}$ for $0 \leq x \leq \pi$ on the same plot. You might like to read Section 6.1.5 on page 179 to learn how to plot multiple curves on the same graph. [Hint: You can use `plot(x,y,x,z,'--')` or you can plot the first curve, use the `hold on` command, and then plot the second curve on top of the first one.]

8. **Fancy plots:** Go to Section 6.1.6 and look at the examples of specialized 2-D plots given there. Reproduce any of the plots you like.

9. **A very difficult plot:** Use your knowledge of *splines* and *interpolation* to draw a lizard (just kidding).

Answers to Exercises

Commands required to solve the problems are shown here.

1. ```
   x=linspace(0,2*pi,100);
   plot(x,sin(x))
   xlabel('x'), ylabel('sin(x)')
   title('Plot created by Rudra Pratap')
   ```

2. ```
   plot(x,sin(x),x,sin(x),'o')
   xlabel('x'), ylabel('sin(x)')
   ```

3. ```
   x=linspace(0,4*pi,10);  % with 10 points
   y=exp(-.4*x).*sin(x);
   plot(x,y)
   x=linspace(0,4*pi,50);  % with 50 points
   y=exp(-.4*x).*sin(x);
   plot(x,y)
   x=linspace(0,4*pi,100); % with 100 points
   y=exp(-.4*x).*sin(x);
   plot(x,y)
   ```

4. ```
   t=linspace(0,20,100);
   plot3(sin(t),cos(t),t)
   ```

5. You should not be looking for an answer here.

6. ```
   x=0:10:1000;
   y=x.^3;
   semilogx(x,y)
   semilogy(x,y)
   loglog(x,y)
   ```

7. ```
   x=linspace(0,pi,100);
   y=cos(x); z=1-x.^2/2+x.^4/24;
   plot(x,y,x,z)
   plot(x,y,x,z,'--')
   legend('cos(x)','z')    % try this legend command
   ```

[For fun: If the last command **legend** does produce a legend on your plot, click and hold your mouse on the legend and see if you can move it to a location of your liking. See page 177 for more information on **legend**.]

2.4 Lesson 4: Creating, Saving, and Executing a Script File

Goal: To learn how to create script files and execute them in MATLAB.

A *script file* is a user-created file with a sequence of MATLAB commands in it. The file must be saved with a .m extension to its name, thereby, making it an *M-file*. A script file is executed by typing its name (without the .m extension) at the command prompt. For more information, see Section 4.1 on page 99.

Time Estimates
Lesson: 20 minutes
Exercises: 30 minutes

What you are going to learn
- How to create, write, and save a script file.
- How to execute the script file in MATLAB.

Unfortunately, creating, editing, and saving files are somewhat system-dependent tasks. The commands needed to accomplish these tasks depend on the operating system and the text editor you use. It is not possible to provide an introduction to these topics here. So, we assume that

- You know how to use a text editor on your UNIX system (for example, *vi* or *emacs*), or that you're using the built-in MATLAB editor on a Mac or a PC.
- You know how to open, edit, and save a file.
- You know which directory your file is saved in.

Method: Write a script file to draw the unit circle of Lesson 3. You are essentially going to write the commands shown in Fig. 2.3 in a file, save it, name it, and execute it in MATLAB. Follow the directions below.

1. Create a new file:
   - **On PCs and Macs:** Select File→New→Blank M-File from the File menu. A new edit window should appear.
   - **On UNIX workstations:** Type !vi circle.m or !emacs circle.m at the MATLAB prompt to open an edit window in *vi* or *emacs*.

2. Type the following lines into this file. Lines starting with a % sign are interpreted as comment lines by MATLAB and are ignored.

```
% CIRCLE - A script file to draw a unit circle
% File written by Rudra Pratap. Last modified 5/28/98
% ------------------------
theta=linspace(0,2*pi,100);    % create vector theta
x=cos(theta);                  % generate x-coordinates
y=sin(theta);                  % generate y-coordinates
plot(x,y);                     % plot the circle
axis('equal');                 % set equal scale on axes
title('Circle of unit radius') % put a title
```

3. Write and save the file under the name circle.m:

   - **On PCs:** Select Save As... from the File menu. A dialog box should appear. Type `circle.m` as the name of the document. Make sure the file is being saved in the folder you want it to be in (the current working folder/directory of MATLAB). Click Save to save the file.

   - **On UNIX workstations:** You can use your favorite editor to write and save the file. After writing the file, quit the editor to get back to MATLAB.

4. Now get back to MATLAB and type the commands shown in Fig. 2.4 in the command window to execute the script file.

```
>> help circle                          Seek help on the script file to see
                                        if MATLAB can access it.


CIRCLE - A script file to draw a unit circle
File written by Rudra Pratap. Last modified 6/28/98.
- - - - - - - - - - - - - - - - - - - - - - -
                                        MATLAB lists the comment lines
                                        of the file as on-line help.

>> circle
                                        Execute the file. You should see
                                        the circle plot in the figure window.
```

Figure 2.4: Executing a script file.

5. If you have the script file open in the MATLAB editor window, you can also execute the file by pressing the **Run** *file* icon (the little green arrowhead over a white page in the edit window nemu bar) or the F5 function key (see **Debug** menu).

EXERCISES

1. **Show the center of the circle:** Modify the script file circle.m to show the center of the circle on the plot, too. Show the center point with a "+". (Hint: See Exercises 2 and 7 of Lesson 3.)

2. **Change the radius of the circle:** Modify the script file circle.m to draw a circle of arbitrary radius r as follows:
   - Include the following command in the script file before the first executable line (theta=...) to ask the user to input (r) on the screen:
     ```
     r = input('Enter the radius of the circle:  ')
     ```
   - Modify the x- and y-coordinate calculations appropriately.
   - Save and execute the file. When asked, enter a value for the radius and press return.

3. **Variables in the workspace:** All variables created by a script file are left in the global workspace. You can get information about them and access them, too:
   - Type who to see the variables present in your workspace. You should see the variables r, theta, x, and y in the list.
   - Type whos to get more information about the variables and the workspace.
   - Type [theta' x' y'] to see the values of θ, x, and y listed as three columns. All three variables are row vectors. Typing a single right quote (') after their names transposes them and makes them column vectors.

4. **Contents of the file:** You can see the contents of an M-file without opening the file with an editor. The contents are displayed by the type command. To see the contents of circle.m, type type circle.m.

5. **H1 line:** The first commented line before any executable statement in a script file is called the *H1 line*. It is this line that is searched by the lookfor command. Because the lookfor command is used to look for M-files with keywords in their description, you should put keywords in the H1 line of all M-files you create. Type lookfor unit to see what MATLAB comes up with. Does it list the script file you just created?

6. **Just for fun:** Write a script file that, when executed, greets you, displays the date and time, and curses your favorite TA or professor. [The commands you need are disp, date, clock, and possibly fix. See the on-line help on these commands before using them.]

Answers to Exercises

1. Replace the command plot(x,y) by the command plot(x,y,0,0,'+').

2. Your changed script file should look like this:

```
% CIRCLE - A script file to draw a circle of radius 'r'
% Original circle.m file modified to incorporate variable r
% File written by Rudra Pratap on 9/14/94.
%                        Last modified 5/28/2009
% ---------------------------
r=input('Enter the radius of the circle: ')
theta=linspace(0,2*pi,100);          % create vector theta
x=r*cos(theta);                      % generate x-coordinates
y=r*sin(theta);                      % generate y-coordinates
plot(x,y);                           % plot the circle
axis('equal');                       % set equal scale on axes
title('Circle of given radius r')    % put a title
```

6. Here is a script file that you may not fully understand yet. Do not worry, just copy it if you like it. See the on-line help on the commands used, e.g., disp, date, fix, clock, and int2str.

```
% Script file to begin your day. Save it as Hi_there.m
% To execute, just type Hi_there
% File written by Rudra Pratap on 6/15/95
%                        Last modified 6/28/98
% --------------------------
disp('Hello R.P., How is life?')
disp(' ')                            % display a blank line
disp('Today is...')
disp(date)                           % display date
time=fix(clock);                     % get time as integers
hourstr=int2str(time(4));            % get the hour
minstr=int2str(time(5));             % get the minute
if time(5)<10                        % if minute is, say 5, then
    minstr=['0',minstr];             %- write it as 05
end
timex=[hourstr ':' minstr];          % create the time string
disp(' ')
disp('And the time is..')
disp(timex)                          % display the time
```

2.5   Lesson 5: Creating and Executing a Function File

Goal: To learn how to write and execute a function file and to learn the difference between a script file and a function file.

A *function file* is also an M-file, just like a script file, except it has a function definition line on the top that defines the input and output explicitly. For more information, see Section 4.2.

Time Estimates
Lesson: 15 minutes
Exercises: 60 minutes

What you are going to learn

- How to open and edit an existing M-file.

- How to define and execute a function file.

Method: Write a function file to draw a circle of a specified radius, with the radius as the input to the function. You can either write the function file from scratch or modify the script file of Lesson 4. We advise you to select the latter option.

1. Open the script file circle.m:

    - **On PCs:** Select File→Open... from the File menu. Navigate and select the file circle.m from the Open dialog box. Double-click to open the file. The contents of the file should appear in an edit window.

    - **On UNIX workstations:** Type !vi circle.m or !emacs circle.m on the MATLAB prompt to open the file in a *vi* or *emacs* window.

2. Edit the file circle.m from Lesson 4 to look like the following:

    ```
    function [x,y] = circlefn(r);
    % CIRCLEFN - Function to draw a circle of radius r.
    % File written by Rudra Pratap on 9/17/94. Last modified 7/1/98
    % Call syntax:   [x,y] = circlefn(r);  or just:  circlefn(r);
    % Input:      r = specified radius
    % Output: [x,y] = the x- and y-coordinates of data points
    theta=linspace(0,2*pi,100);   % create vector theta
    x = r*cos(theta);             % generate x-coordinates
    y = r*sin(theta);             % generate y-coordinates
    plot(x,y);                    % plot the circle
    axis('equal');                % set equal scale on axes
    title(['Circle of radius r =',num2str(r)])
                                  % put a title with the value of r
    ```

Alternatively, you could select File→New→Function M-File and type all the lines in the new file.

3. Now write and save the file under the name circlefn.m as follows:

- **On PCs:** Select **Save As...** from the **File** menu. A dialog box should appear. Type `circlefn.m` as the name of the document (usually, MATLAB automatically writes the name of the function in the document name). Make sure the file is saved in the folder you want (the current working folder/directory of MATLAB). Click **Save** to save the file.

- **On UNIX workstations:** You are on your own to write and save the file using the editor of your choice. After writing the file, quit the editor to get back to MATLAB.

4. Figure 2.5 shows a sample session that executes the function `circlefn` in three different ways. Try it out.

`>> R = 5;`	Specify the input and execute the function with an explicit output list.
`>> [x,y] = circlefn(R);`	
`>> [cx,cy] = circlefn(2.5);`	You can also specify the value of the input directly.
`>> circlefn(1);`	If you don't need the output, you don't have to specify it.
`>> circlefn(R^2/(R+5*sin(R)));`	Of course, the input can also be a valid MATLAB expression.

Figure 2.5: Lesson 5: Executing a function file.

Comments:

- Note that a function file (see previous page) must begin with a function definition line. To learn more about function files, refer to Section 4.2 on page 102.
- The argument of the `title` command in this function file is slightly complicated. It consists of two character strings. The first one is a simple character string, `'Circle of radius r ='`. The second one, `num2str(r)`, is a function that converts the numeric value of r to a string (and hence the name of the function). The square brackets create an array of the two strings by concatenating them. You will learn more about this in Section 3.3 on page 77.

EXERCISES

1. **On-line help:** Type `help function` to get on-line help on `function`. Read through the help file.

2. **Convert temperature:** Write a function that outputs a conversion table for Celsius and Fahrenheit temperatures. The input of the function should be two numbers: T_i and T_f, specifying the lower and upper range of the table in Celsius. The output should be a two-column matrix: the first column showing the temperature in Celsius from T_i to T_f in the increments of 1°C and the second column showing the corresponding temperatures in Fahrenheit. To do this, (i) create a column vector C from T_i to T_f with the command `C=[Ti:Tf]'`, (ii) calculate the corresponding numbers in Fahrenheit using the formula $[F = \frac{9}{5}C + 32]$, and (iii) make the final matrix with the command `temp=[C F];`. Note that your output variable will be named *temp*.

3. **Calculate factorials:** Write a function `factorial` to compute the factorial $n!$ for any integer n. The input should be the number n and the output should be $n!$. You might have to use a *for* loop or a *while* loop to do the calculation. See Section 4.3.4 on page 114 for a quick description of these loops. (You can use the built-in function `prod` to calculate factorials. For example, $n!$ is `prod(1:n)`. In this exercise, however, do not use this function.)

4. **Compute the cross product:** Write a function file `crossprod` to compute the cross product of two vectors $\mathbf{u}$ and $\mathbf{v}$, given $\mathbf{u} = (u_1, u_2, u_3)$, $\mathbf{v} = (v_1, v_2, v_3)$, and $\mathbf{u} \times \mathbf{v} = (u_2v_3 - u_3v_2, u_3v_1 - u_1v_3, u_1v_2 - u_2v_1)$. Check your function by taking cross products of pairs of unit vectors: $(\mathbf{i}, \mathbf{j})$, $(\mathbf{j}, \mathbf{k})$, etc. $[\mathbf{i} = (1, 0, 0), \ \mathbf{j} = (0, 1, 0), \ \mathbf{k} = (0, 0, 1)]$. (Do not use the built-in function `cross` here.)

5. **Sum a geometric series:** Write a function to compute the sum of a geometric series $1 + r + r^2 + r^3 + \cdots + r^n$ for a given r and n. Thus the input to the function must be r and n and the output must be the sum of the series. [See Exercise 4 of Lesson 2.]

6. **Calculate the interest on your money:** The interest you get at the end of n years, at a flat annual rate of $r\%$, depends on how the interest is compounded. If the interest is added to your account k times a year, and the principal amount you invested is X_0, then at the end of n years you would have $X = X_0 \left(1 + \frac{r}{k}\right)^{kn}$ amount of money in your account. Write a function to compute the interest $(X - X_0)$ on your account for a given X, n, r, and k. Use the function to find the difference between the interest paid on $1000 at the rate of 6% a year at the end of five years if the interest is compounded (i) quarterly ($k = 4$) and (ii) daily ($k = 365$). For screen output, use `format bank` (see Section 1.6.3, page 9, for a description of various formats).

Answers to Exercises

Some of the commands in the following functions might be too advanced for you at this point. If so, look them up or ignore them for now.

```
2. function temptable = ctof(tinitial,tfinal);
   % CTOF : function to convert temperature from C to F
   % call syntax:
   %          temptable = ctof(tinitial,tfinal);
   % ------------
   C = [tinitial:tfinal]';        % create a column vector C
   F = (9/5)*C + 32;              % compute corresponding F
   temptable = [C F];             % make a 2 column matrix of C & F
```

Note: Once the function file is written and saved, you could run it to get a temperature table showing conversion from, say 0 to 100 degrees Celsius. You can either use ctof(0,100) or table = ctof(0,100) to get the table. Now, here is a note of caution. What happens if you type ctof([0,100]) instead of ctof(0,100)? You will get an error message. The two commands are very different although they differ just by a pair of square brackets. The function, as written, expects two numbers in the input list, *tinitial* and *tfinal*. Therefore, when you type ctof(0,100), the function automatically sets tinial = 0 and tfinal = 100. The computation proceeds accordingly. However, when you type ctof([0,100]), then the input is a single *vector* [0,100] that has two elements: 0 and 100. The function, however, is not written to accept a single vector as its input and, therefore, the function gets confused and produces an error message. The moral of the story is, there is a difference between a list of numbers and a vector or an array as a single object. This is something that you will have to master as you work in MATLAB.

```
3. function factn = factorial(n);
   % FACTORIAL: function to compute factorial n!
   % call syntax:
   %          factn = factorial(n);
   % ------------
   factn = 1;                     % initialize. also 0! = 1
   for k = n:-1:1                 % go from n to 1
       factn = factn*k;           % multiply n by n-1, n-2, etc.
   end
```

Can you modify this function to check for negative input and noninteger input before it computes the factorial?

```
4. function w = crossprod(u,v);
   % CROSSPROD: function to compute w = u x v for vectors u & v
   % call syntax:
   %          w = crossprod(u,v);
   % ------------
   if length(u)>3 | length(v)>3  % check if u OR v has > 3 elements
      error('Ask Euler. This cross product is beyond me.')
   end
   w = [u(2)*v(3)-u(3)*v(2);     % first element of w
        u(3)*v(1)-u(1)*v(3);     % second element of w
        u(1)*v(2)-u(2)*v(1)];    % third element of w
```

Can you modify this function to check for 2-D vectors and set the third component automatically to zero?

```
5. function s = gseriessum(r,n);
   % GSERIESSUM: function to calculate the sum of a geometric series
   % The series is  1+r+r^2+r^3+....r^n (up to nth power).
   % call syntax:
   %          s = gseriessum(r,n);
   % ------------
   nvector = 0:n;                % create a vector from 0 to n
   series = r.^nvector;          % create a vector of terms in the series
   s = sum(series);              % sum all elements of the vector 'series'.
```

```
6. function [capital,interest] = compound(capital,years,rate,timescomp);
   % COMPOUND: function to compute the compounded capital and the interest
   % call syntax:
   %          [capital,interest] = compound(capital,years,rate,timescomp);
   % ------------
   x0 = capital; n = years; r = rate; k = timescomp;
   if r>1                        % check for common mistake
      disp('check your interest rate. For 8% enter .08, not 8.')
   end
   capital = x0*(1+r/k)^(k*n);
   interest = capital - x0;
```

[Ans. (i) Quarterly: $346.86,   Daily: $349.83,   Difference: $2.97.]

Directional Explorations

Here are the next set of lessons. These lessons are a little longer than the previous ones and rely upon the basic skills you have already developed with the previous lessons.

Lesson 6: Manipulate matrices and use them as matrices or arrays.
Key features: Learn how to use arrays efficiently, manipulate rows, columns, or chunks of matrices and distinguish between matrix and array computation.

Lesson 7: Create and work with anonymous functions.
Key features: Learn how to create functions on-the-fly (without using files) and use them in computation efficiently.

Lesson 8: Work with symbolic mathematics toolbox.
Key features: Create symbolic variables, do symbolic algebra and manipulation, and use symbolic math functions.

Lesson 9: Saving, loading, importing, and exporting data.
Key features: Learn how to save computed data in a file, load the data back into the workspace, and read different kinds of data files.

Lesson 10: Learn about file and directory navigation.
Key features: Learn several ways of checking your current directory, changing working directory, and setting MATLAB path.

Lesson 11: Generate report from your MATLAB programs using *publisher*.
Key features: Learn how to generate reports in HTML format from your MATLAB work.

These lessons use many new commands. We strongly recommend that whenever you use a new command that is not explained, use on-line help to first read about the command. Many commands are not explained in detail because their descriptions are readily available as on-line help.

Another difference with the previous lessons is that answers to exercises are not provided. In most cases, answers are not required. However, solutions to these exercises are available on the author's website (along with additional material).

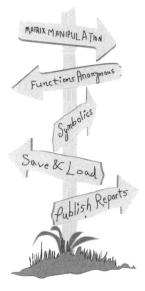

2.6   Lesson 6: Working with Arrays and Matrices

Goal: To become familiar with 2-D arrays, indexing, matrix manipulation and simple computations with arrays and matrices.

Time Estimates
Lesson: 20 minutes
Exercises: 30 minutes

What you are going to learn

- How to enter a matrix.
- How to access an element, a row, a column, or a submatrix of a matrix.
- How to multiply a matrix with a vector.
- How to distinguish between array operation and matrix operation.

Method: We will take a simple 3×3 matrix and use its row and columns indices to manipulate the matrix, extract a vector, compute inner and outer product of vectors, multiply the matrix with a vector, exponentiate the matrix, and exponentiating each element of the matrix using an *array operation*.

Comments: Arrays are the backbone of MATLAB computation. The more familiar you are with arrays, their manipulation and usage (Figs. 2.6 and 2.7), the better off you are in the world of MATLAB. You should look out for understanding the following two concepts and acquiring associated computational skills.

- A 2-D array is a list of numbers arranged in rows and columns. If you form an array by writing numbers in rows, all rows must have the same number of entries. Same is true for columns. An array with m rows and n columns is called an $m \times n$ array and it has a total of $m \cdot n$ entries. An element of the array is recognized by its location—its row number and column number. These row and column identifiers are called indices of the matrix. Thus $A(i, j)$ refers to a specific element of matrix A located in the ith row and jth column. These locations are unique. In MATLAB, you can also use a list of numbers for the row and column indices to access more than one element of the matrix at a time. This concept of using indices carries over to higher dimensional arrays (discussed in Section 4.4) as well.

- You can use arrays for carrying out element-by-element operations (also called *array operations*, see Section 3.2 for details) as well as matrix operations. If A is a square array, then $B = [A] * [A]$ or $B = A^2$ is very different from $B = [a_{ij}^2]$. The first one is the square of matrix A or the product of the matrix A with itself, whereas the second one is an array made of squares of corresponding elements of array A.

```
>> A=[1 2 3; 4 5 6; 7 8 8]

A =

       1       2       3
       4       5       6
       7       8       8
```

Matrices are entered row-wise.
Rows are separated by semicolons
and columns are separated by
spaces or commas.

```
>> A(2,3)

ans =

       6
```

Element A_{ij} of matrix A is
accessed as A(i,j).

```
>> A(3,3) = 9

A =

       1       2       3
       4       5       6
       7       8       9
```

Correcting any entry is easy
through indexing.

```
>> B = A(2:3,1:3)

B =

       4       5       6
       7       8       9
```

Any submatrix of A is obtained
by using range specifiers for row
and column indices.

```
>> B = A(2:3,:)

B =

       4       5       6
       7       8       9
```

The colon by itself as a row or
column index specifies all rows or
columns of the matrix.

```
>> B(:,2)=[]

B =

       4       6
       7       9
```

A row or a column of a matrix is
deleted by setting it to a null
vector [].

Figure 2.6: Lesson 6: Examples of matrix input and matrix manipulation.

```
>> A = [1 2 3; 4 5 6; 7 8 9];
>> x = A(1,:)'

x =
     1
     2
     3
```

Matrices are transposed using the single right-quote character ('). Here x is the transpose of the first row of A.

```
>> x'*x

ans =
    14
```

Matrix or vector products are well-defined between compatible pairs. A row vector (x') times a column vector (x) of the same length gives the inner product, which is a scalar, but a column vector times a row vector of the same length gives the outer product, which is a matrix.

```
>> x*x'

ans =
     1     2     3
     2     4     6
     3     6     9
```

```
>> A*x

ans =
    14
    32
    50
```

Look how easy it is to multiply a vector with a matrix, compared with Fortran or Pascal.

```
>> A^2

ans =
    30    36    42
    66    81    96
   102   126   150
```

You can even exponentiate a matrix if it is a square matrix. $A\verb|^|2$ is simply $A*A$.

```
>> A.^2

ans =
     1     4     9
    16    25    36
    49    64    81
```

When a dot precedes the arithmetic operators *, ^, and /, MATLAB performs array operation (element-by-element operation). So, A.^2 produces a matrix with elements (a_{ij})^2.

Figure 2.7: Lesson 6: Examples of matrix transpose, matrix multiplication, matrix exponentiation, and array exponentiation.

EXERCISES

✳ 1. **Entering matrices:**   Enter the following three matrices.

$$A = \begin{bmatrix} 2 & 6 \\ 3 & 9 \end{bmatrix}, \qquad B = \begin{bmatrix} 1 & 2 \\ 3 & 4 \end{bmatrix}, \qquad C = \begin{bmatrix} -5 & 5 \\ 5 & 3 \end{bmatrix}.$$

2. **Check some linear algebra rules:**

   - **Is matrix addition commutative?** Compute A+B and then B+A. Are the results the same?
   - **Is matrix addition associative?** Compute (A+B)+C and then A+(B+C) in the order prescribed. Are the results the same?
   - **Is multiplication with a scalar distributive?** Compute α(A+B) and αA+αB, taking $\alpha = 5$, and show that the results are the same.
   - **Is multiplication with a matrix distributive?** Compute A*(B+C) and compare with A*B+A*C.
   - **Matrices are different from scalars!** For scalars, $ab = ac$ implies that $b = c$ if $a \neq 0$. Is that true for matrices? Check by computing A*B and A*C for the matrices given in Exercise 1. Also, show that A*B $\neq$ B*A.

✳ 3. **Create matrices with zeros, eye, and ones:** Create the following matrices with the help of the matrix generation functions zeros, eye, and ones. See the on-line help on these functions, if required (e.g., help eye).

$$D = \begin{bmatrix} 0 & 0 & 0 \\ 0 & 0 & 0 \end{bmatrix}, \qquad E = \begin{bmatrix} 5 & 0 & 0 \\ 0 & 5 & 0 \\ 0 & 0 & 5 \end{bmatrix}, \qquad F = \begin{bmatrix} 3 & 3 \\ 3 & 3 \end{bmatrix}.$$

✳ 4. **Create a big matrix with submatrices:** The following matrix G is created by putting matrices A, B, and C, given previously, on its diagonal. In how many ways can you create this matrix using submatrices A, B, and C (that is, you are not allowed to enter the nonzero numbers explicitly)?

$$G = \begin{bmatrix} 2 & 6 & 0 & 0 & 0 & 0 \\ 3 & 9 & 0 & 0 & 0 & 0 \\ 0 & 0 & 1 & 2 & 0 & 0 \\ 0 & 0 & 3 & 4 & 0 & 0 \\ 0 & 0 & 0 & 0 & -5 & 5 \\ 0 & 0 & 0 & 0 & 5 & 3 \end{bmatrix}.$$

5. **Manipulate a matrix:** Do the following operations on matrix G created in Exercise 4.

   - Delete the last row and last column of the matrix.
   - Extract the first 4×4 submatrix from G.
   - Replace $G(5,5)$ with 4.
   - What do you get if you type G(13) and hit return? Can you explain how MATLAB got that answer?
   - What happens if you type G(12,1)=1 and hit return?

2.7   Lesson 7: Working with Anonymous Functions

Goal: To learn how to define and use anonymous functions in command-line computations.

An anonymous function is a function of one or more variables that you create on the command line for subsequent evaluation (Figs 2.8 and 2.9). Such a function is especially useful if you need to evaluate the function several times (with different input) during a single MATLAB session and you do not care to code it in a function file and save for later use. For example, let us say that you have $f(x) = x^3 - 32x^2 + (x - 22)x + 100$, and you need to evaluate this function at different values of x, plot it over certain range of x, or find the zeros of $f(x)$. For such computations, you can define $f(x)$ as an anonymous function and evaluate it at any value(s) of x or use this function as an input to other functions that need such input.

Time Estimates
Lesson: 25 minutes
Exercises: 40 minutes

What you are going to learn

- How to define an anonymous function of a single variable.
- How to evaluate an anonymous function with scalar or array arguments.
- How to define an anonymous function of several variables.
- How to use anonymous functions as input to other functions.

Method:
The key to anonymous functions is the syntax of its creation:

fn_name = @(*list of input variables*) *function_expression*   .

Here, the key element is the symbol @ that assigns a *function handle* to the defined function. A *function handle* is a name given to a function by which you can call it wherever you need it. In the anonymous function definition line, *fn_name* is the name of the function or the *handle* of the function. The syntax @(*list of input variables*) is what tells MATLAB that you are defining an anonymous function here.

Comments:

- Anonymous functions are defined on the command line. They live in the MATLAB workspace and are alive as long as the MATLAB session lasts.
- You can define an anonymous function with any number of input variables.
- You must use a vectorized expression (using array operators) for the function if you intend to use an array as an input variable.
- You can use anonymous functions as input to other functions where appropriate.

```
>> f = @(x) x^3-3*x^2 +x*log(x-1)+100
```

f =

Create a function
$$f(x) = x^3 - 3x^2 + x\log(x-1) + 100$$

```
   @(x)x^3-3*x^2+x*log(x-1)+100
```

```
>> f(0)
```

Evaluate the function at $x = 0$, i.e., find $f(0)$.

ans =

```
   100
```

```
>> f(1)
```

Evaluate the function at $x = 1$. Note that f is singular at $x = 1$.

ans =

```
  -Inf
```

```
>> values = [f(0)  f(1)  f(2)  f(10)]
```

values =

You can use f in an array also.

```
   100.0000      -Inf   96.0000   821.9722
```

```
>> x=[0 1 2 10];
>> f(x)
???? Error using ==> mpower
Matrix must be square.
```

Using an array as the input to f causes an error. This is because the expression for f is not vectorized.

```
Error in ==> @(x)x^3-3*x^2+x*log(x-1)+100
```

```
>> f = @(x) x.^3-3*x.^2 +x.*log(x-1)+100;
f(x)
```

ans =

Redefine f by vectorizing the expression (use array operators). Now use it with an array argument.

```
   100.0000      -Inf   96.0000   821.9722
```

```
>> x = linspace(-10,10);
>> plot(x,f(x))
```

You can also use f as input to other functions where appropriate.

```
Warning: Imaginary parts of complex X and/or Y
arguments ignored
```

Figure 2.8: Lesson 7: Creating and evaluating anonymous functions in one variable.

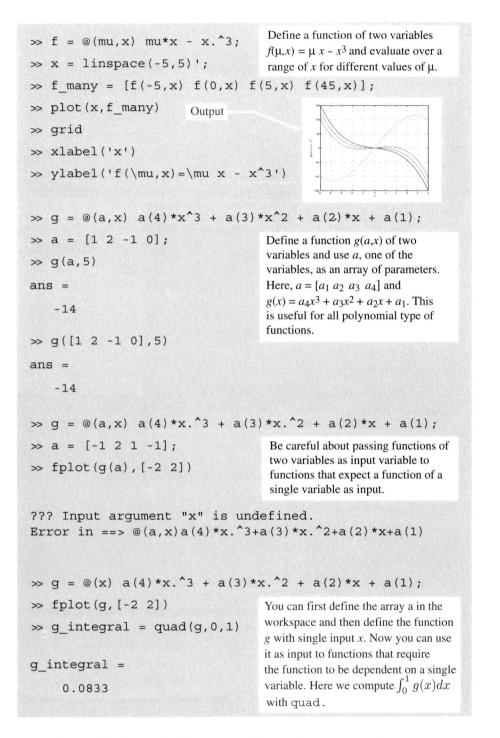

```
>> f = @(mu,x) mu*x - x.^3;
```
Define a function of two variables
$f(\mu,x) = \mu\, x - x^3$ and evaluate over a
range of x for different values of μ.
```
>> x = linspace(-5,5)';
>> f_many = [f(-5,x) f(0,x) f(5,x) f(45,x)];
>> plot(x,f_many)
>> grid
>> xlabel('x')
>> ylabel('f(\mu,x)=\mu x - x^3')
```

Output

```
>> g = @(a,x) a(4)*x^3 + a(3)*x^2 + a(2)*x + a(1);
>> a = [1 2 -1 0];
>> g(a,5)

ans =

    -14

>> g([1 2 -1 0],5)

ans =

    -14
```
Define a function $g(a,x)$ of two
variables and use a, one of the
variables, as an array of parameters.
Here, $a = [a_1\ a_2\ a_3\ a_4]$ and
$g(x) = a_4 x^3 + a_3 x^2 + a_2 x + a_1$. This
is useful for all polynomial type of
functions.

```
>> g = @(a,x) a(4)*x.^3 + a(3)*x.^2 + a(2)*x + a(1);
>> a = [-1 2 1 -1];
>> fplot(g(a),[-2 2])
```
Be careful about passing functions of
two variables as input variable to
functions that expect a function of a
single variable as input.

```
??? Input argument "x" is undefined.
Error in ==> @(a,x)a(4)*x.^3+a(3)*x.^2+a(2)*x+a(1)
```

```
>> g = @(x) a(4)*x.^3 + a(3)*x.^2 + a(2)*x + a(1);
>> fplot(g,[-2 2])
>> g_integral = quad(g,0,1)


g_integral =

    0.0833
```
You can first define the array a in the
workspace and then define the function
g with single input x. Now you can use
it as input to functions that require
the function to be dependent on a single
variable. Here we compute $\int_0^1 g(x)dx$
with quad.

Figure 2.9: Lesson 7: More explorations with anonymous functions.

EXERCISES

✳ 1. **Creation and evaluation of anonymous functions:** Create the function:
$f(x) = x^2 - \sin(x) + \frac{1}{x}$.

   (a) Find $f(0), f(1)$, and $f(\pi/2)$.
   (b) Vectorize f and evaluate $f(x)$ where $x = [\ 0\ \ 1\ \ \pi/2\ \ \pi\ ]$.
   (c) Create x = `linspace(-1,1)`, evaluate $f(x)$, and plot x vs $f(x)$.
   (d) Combine the following three commands into a single command to produce
       the plot that you will get at the end of the third command.

   ```
   x = linspace(-1, 1);    y = f(x);    plot(x,y);
   ```

   (e) Use `fplot` to graph $f(x)$ over x from $-\pi$ to π.

2. **Computation with multiple anonymous functions:** Create three anonymous functions corresponding to the following expressions:

$$
\begin{aligned}
f(x) &= x^4 - 8x^3 + 17x^2 - 4x - 20 \\
g(x) &= x^2 - 4x + 4 \\
h(x) &= x^2 - 4x - 5.
\end{aligned}
$$

   (a) Evaluate $f(x) - g(x)h(x)$ at $x = 3$.
   (b) Evaluate $f(x) - g(x)h(x)$ at $x = [1\ 2\ 3\ 4\ 5]$.
   (c) Evaluate $\frac{f(x)}{g(x)} - h(x)$ for any x.
   (d) Plot $f(x)$ and $\frac{f(x)}{g(x)h(x)}$ over $x \in [-5,5]$.

3. **Anonymous functions as the input to other functions:** Use the same
   function $f(x)$ defined previously, i.e., $f(x) = x^4 - 8x^3 + 17x^2 - 4x - 20$ for
   the following tasks.

   (a) Plot the function using `fplot` over an appropriate interval of x and locate
       the zeros of the function on the graph (that is, find all x for which
       $f(x) = 0$).
   (b) Learn about the built-in function `fzero` that finds the zero of any given
       function by typing `help fzero`. Now use `fzero` to find the zeros of f
       accurately, making approximate initial guesses based on the plot you
       made above.
   (c) Use function `quad` to integrate $f(x)$ from $x = 0$ to $x = 1$ and verify the
       results by direct integration.

✳ 4. **Anonymous functions of several variables:** The formula for computing
   compound interest on an investment is given by

$$
x = x_0 \left(1 + \frac{r}{100}\right)^n
$$

   where x = accumulated amount, x_0 = initial investment, r = rate of annual
   interest in percentage, and n = number of years. Define an anonymous function to compute x with (x_0, r, n) as the input. Using this function, compare
   the growth of a $1000 investment over a period of eight years earning an
   interest of 9% with that over a period of nine years earning an interest of 8%.

2.8   Lesson 8: Symbolic Computation

> This lesson requires that you have Symbolic Math Toolbox installed on your computer in addition to the standard MATLAB. The student version of MATLAB includes the Symbolic Math Toolbox. Before proceeding, other users must check (click on Help → Product Help or the question mark (?) on the menu bar to list installed MATLAB-related products) to see if they have access to this toolbox.

Goal: To learn how to do simple symbolic algebra in MATLAB. Such calculations are done with symbolic variables without resorting to their numerical values.

Time Estimates

Lesson: 20 minutes
Exercises: 60 minutes

What you are going to learn

- How to create symbolic variables and use them in defining functions.
- How to manipulate expressions using symbolic math functions `expand`, `simplify`, `subs`, `factor`, `pretty`, etc.
- How to differentiate and integrate functions symbolically.
- How to solve simultaneous linear and nonlinear equations.

Method: The most important step in carrying out symbolic computation is to declare the independent variables to be symbolic before you do anything with them. Suppose you want to use x and y as symbolic variables. Then, you can declare them to be symbolic using any of the commands that follow:

- `x = sym('x');    y = sym('y')`
- `syms x y`

The command `sym` is the formal command for constructing objects of symbol class. You can define one symbolic variable at a time using this command. You can use additional qualifiers in the definition for the symbolic object (e.g., for restricting a variable to be only *real* or *positive*). The command `syms` is a shortcut to declare several variables of the same class in one shot. We will use `syms` in this lesson. For learning more about these commands and symbolic computation, see Chapter 8.

There are three basic skills to learn in symbolic computations: (i) How to define expressions and do simple algebra with them (multiply, divide, expand, factorize, simplify, etc.). (ii) How to do simple calculus (differentiate, integrate, etc.). (iii) How to solve equations—algebraic and differential equations. Once you know these skills (Figs. 2.10 and 2.11), you can meet most of your symbolic math calculation requirements. The rest of it is about honing these skills, learning clever shortcuts and substitutions, and establishing a personal rapport with the Symbolic Math Toolbox for having fun with these calculations.

```
>> syms x y
>> f = (x+y)^3
f =
(x + y)^3
```

Declare *x* and *y* to be symbolic variables. Define a function *f* as
$$f = (x + y)^3.$$

```
>> expand(f)
ans =
x^3 + 3*x^2*y + 3*x*y^2 + y^3
```

Use expand to multiply out and expand algebraic or trigonometric expressions.

```
>> factor(ans)
ans =
(x + y)^3
```

Use factor to find factors of long algebraic expressions.

```
>> z = sin(x+y);
>> expand(z)
ans =
cos(x)*sin(y) + cos(y)*sin(x)
```

Define a trigonometric expression.

```
>> subs(z, y, pi-x)
ans =
0
```

Substitute $y = \pi{-}x$ in expression *z*.

```
>> diff(z,x)
ans =
cos(x + y)
>> z_xx = diff(z,x,2)
z_xx =
-sin(x + y)
```

Differentiate *z* with respect to *x*, i.e., find $\frac{\partial z}{\partial x}$.

Find the second derivative of *z* with respect to *x*, i.e., find $\frac{\partial^2 z}{\partial x^2}$.

```
>> int(z,x,0,pi/2)
ans =
cos(y) + sin(y)
```

Integrate *z* with respect to *x* from 0 to π/2, i.e., evaluate $\int_0^{\frac{\pi}{2}} z\,dx$.

Figure 2.10: Lesson 8: Symbolic calculations.

```
>> v = [x; y];

>> inner_product = v'*v

inner_product =

x*conj(x) + y*conj(y)


>> syms x y real

>> inner_product

inner_product =

x^2 + y^2


>> syms a b

>> exp1 = 'a*x + b*y -3';

>> exp2 = '-x + 2*a*y -5';

>> [x,y] = solve(exp1, exp2)

x =

(6*a - 5*b)/(2*a^2 + b)

y =

(5*a + 3)/(2*a^2 + b)


>> subs(exp1)

ans =

(b*(5*a + 3))/(2*a^2 + b)+(a*(6*a - 5*b))/(2*a^2 + b)-3


>> simplify(ans)

ans =

0


>> pretty(subs(exp2))

   2 a (5 a + 3)      6 a - 5 b
   --------------  -  ---------  - 5

         2                 2
      2 a  + b          2 a  + b
```

Define a column vector $v = [x \; y]^T$. Find the inner product $v^T v$. Note that MATLAB assumes x and y to be complex (default data type) variables and hence
$$v^T v = x\bar{x} + y\bar{y} \; .$$

Now declare (redefine) x and y to be real and evaluate the inner product. Since x and y are real,
$$v^T v = x^2 + y^2 \; .$$

Solve two simultaneous algebraic equations for x and y:
$$ax + by - 3 = 0$$
$$-x + 2ay - 5 = 0 \; .$$

Substitute the values of x and y just found in exp1 to check the result.

Simplify the answer to see if it reduces to zero (as it must in order to satisfy the equation).

Use pretty to get the expression in more readable form.

Figure 2.11: Lesson 8: Symbolic calculations.

EXERCISES

* 1. **Some basic symbolic manipulations:** Define the following function symbolically

$$f(x) = (x^2 - 4x)(x^2 - 4x + 1) - 20.$$

   (a) Expand $f(x)$ using expand.
   (b) Show that $(x^2 - 4x + 4)$ is a factor of f by dividing f with this factor.
   (c) Factorize f using the symbolic function factorize and verify the factor you used in (b).
   (d) Find the roots of one of the two factors. (Hint: Take one of the factors as an expression and use solve to find two roots x_1 and x_2).
   (e) Substitute one of the roots for x, say $x = x_1$, in f and simplify the expression to show that $f(x_1) = 0$.

2. **Computation with symbolic vectors:** Define $r = [x \ y]^T$ where x and y are declared as real variables. Compute
   (a) rr^T (the outer product of r).
   (b) $\int_0^1 rr^T dx$.
   (c) $\int_0^1 \int_0^1 rr^T dx dy$.
   (d) Determinant of $\int_0^1 \int_0^1 rr^T dx dy$.

3. **Solving simultaneous linear equations:** Solve the following set of simultaneous linear algebraic equations.

$$\begin{aligned} x + 3y - z &= 2 \\ x - y + z &= 3 \\ 3x - 5y &= 4. \end{aligned}$$

4. **Solving simultaneous linear equations:** Solve the following nonlinear algebraic equations simultaneously.

$$\begin{aligned} 3x^3 + x^2 - 1 &= 0 \\ x^4 - 10x^2 + 2 &= 0. \end{aligned}$$

5. **Integrals:** Find the following integrals:

$$\text{(i)} \int_0^\infty e^{-x} dx, \quad \text{(ii)} \int_0^\infty e^{-x^2} dx, \quad \text{and (iii)} \int_0^\infty e^{-x^2} \sin(x) dx.$$

6. **Solving differential equations:** Use function dsolve to solve the following differential equations along with the given initial conditions.
   (a) $\frac{dx}{dt} + x^2 = 0$, $x(0) = x_0$.
   (b) $\frac{d^2y}{dt^2} + \omega^2 y = 0$, $y(0) = y_0$ and $\frac{dy}{dt}(0) = v_0$.
   [First, use on-line help on dsolve to see how to enter the differential equation and the initial conditions as input to the function. Please note that this function requires the differential operator $\frac{d}{dt}$ to be denoted by D. Therefore, the first differential equation will be written as   Dx + x^2 = 0].

2.9   Lesson 9: Importing and Exporting Data

Goal: To learn how to read data from common data files into MATLAB workspace and how to save data into a MATLAB readable file.

Time Estimates
Lesson: 30 minutes
Exercises: 30 minutes

What you are going to learn
- How to save data (computed variables) into a Mat-file, the MATLAB's native binary format.
- How to read all variables or some selected variables from data stored in a Mat-file.
- How to write input data as text in a script file (an M-file) and run the script file to load the data in MATLAB workspace.
- How to read mixed data—text and numbers—from an Excel spreadsheet and how to decipher the data read automatically into a cell array.

Method: We are going to work with three different kinds of files for reading data into MATLAB's workspace:
- **Mat-file:** This is MATLAB's native binary format file for saving data. Two commands, `save` and `load` make it particularly easy to save data into and load data from these files. In this lesson, we will create such a file and save some data into it in named variables. We will then use this file to load desired data back into the MATLAB workspace (Fig. 2.12).
- **M-file:** If you have a text file containing data, or you want to write a text file containing data that you would eventually like to read in MATLAB, making it an M-file may be an excellent option. We will use the following M-file (a script file) in this lesson. So, you should first create this file and save it as **TempData.m** before you try the session shown in Fig. 2.13.

```
% TempData: Script file containing data on monthly maximum temperature
Sl_No = [1:12]';
Month = char('January','February','March','April','May','June',...
        'July','August','September','October','November','December');
Ave_Tmax = [22 25 30 34 36 30 29 27 24 23 21 20]';
```

- **Microsoft Excel file:** You can import data from a Microsoft Excel spreadsheet into MATLAB. You can use MATLAB's import wizard, invoked by typing `uiimport` on the command prompt, or by clicking on File → Import Data. MATLAB also provides a special function `xlsread` for reading Excel's spreadsheets as .xls files. In this lesson, we will use `xlsread` (see on-line help before using) to read mixed data from an Excel file **TempData.xls**. This file contains column titles in the first row, numeric data in the first and third column, and text data in the second column. You should first create this Excel file before trying out the commands shown in Fig. 2.13.

TempData.xls

SN	Month	Ave. Tmax
1	January	22
2	February	25
3	March	30
4	April	34
5	May	36
6	June	30
7	July	29
8	August	27
9	September	24
10	October	23
11	November	21
12	December	20

```
>> clear all
>> theta=linspace(0,2*pi,201);
>> r = sqrt(abs(2*sin(4*theta)));
>> x = r.*cos(theta);
>> y = r.*sin(theta);
>> f = char('sqrt(abs(2*sin(4*theta)))');
```
Clear the MATLAB workspace
(all variables are deleted).

```
>> save xydata x y f
>> whos
```
Save variables *x*, *y* and *f* in a binary
(Mat) datafile `xydata.mat`. The
file is created by the `save` command.

```
    Name          Size            Bytes   Class        Attributes
     f            1x27               54    char
     r            1x201            1608    double
     theta        1x201            1608    double
     x            1x201            1608    double
     y            1x201            1608    double
```

```
>> clear all
>> whos
>> load xydata
>> whos
```
Clear all variables, query to see no
variables are present, load the
datafile, and query the workspace
again to see the loaded variables.

```
    Name          Size            Bytes   Class        Attributes
     f            1x27               54    char
     x            1x201            1608    double
     y            1x201            1608    double
```

```
>> plot(x,y),axis('square');
>> f

f =

sqrt(abs(2*sin(4*theta)))
```
The newly loaded variables are *f*, *x*
and *y*. Use these variables to verify
the data they contain.

```
>> clear all
>> load xydata x y
>> whos
```
You can also load only selected
variables from the Mat-file.

```
    Name          Size            Bytes   Class        Attributes
     x            1x201            1608    double
     y            1x201            1608    double
```

```
>> clear all
>> %Action in the file browser
>> size(x)

ans =

     1       201
```
Delete all variables. Select `xydata`
in the current directory pane and
double-click *x* in the data browser to
load *x* in the workspace .

Figure 2.12: Lesson 9: Saving and loading data with MATLAB's native binary data
format is the easiest and most reliable. Examples of how to save data in a Mat-file
and how to load all or some variables from a Mat-file are shown here.

```
>> clear all
>> TempData
>> whos
```
Clear the MATLAB workspace.
Execute the *script file* TempData
and check the new variables.

```
  Name            Size        Bytes  Class       Attributes
  Ave_Tmax        12x1           96  double
  Month           12x9          216  char
  Sl_No           12x1           96  double
```

```
>> Month

Month =

January
February
March
April
...
```
Check one of the variables, Month,
to see the data it contains. All data
typed in file TempData is loaded.

Using script files to load data typed in
them is a very safe and sure shot
method of getting data in the
MATLAB workspace.

```
>>[A,txt,raw] = xlsread('TempData.xls');

>> raw

raw =

    'SN'      'Month'        'Ave. Tmax.'
    [ 1]      'January'      [       22]
    [ 2]      'February'     [       25]
    [ 3]      'March'        [       30]
    ...
    [11]      'November'     [       21]
    [12]      'December'     [       20]
```
Read data from an MS Excel file
that contains a header line, text, and
numeric data in three columns.

Note that the output raw
that contains all the data
is a 13 by 3 cell. You need
to work with this cell to get
readily usable data from it.

```
>> sn = cell2mat(raw(2:13,1));
```
Use cell2mat to convert cell data
in raw(2:13,1) into an array sn.

```
>> months = char(raw{2:13,2});
```
Use char to convert text data in the
second column of raw into a text array.

```
>> T = cell2mat(raw(2:13,3));
```
Similarly, get the numeric data in the
third column into array T.

```
>> bar(sn,T)
```

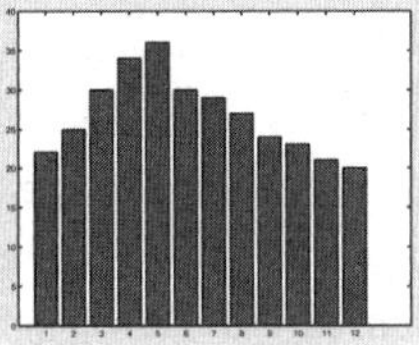

Now you can use the numeric arrays
sn and T as normal data. Here we
use them to create a bar graph
(output shown on the left).

Figure 2.13: Lesson 9: Loading data typed as text in a script file is easy and reliable. You can also import data from other MATLAB readable file formats using File → Import Data. Here, we use function xlsread to load data from an Excel file.

EXERCISES

1. **Working with binary Mat-files:** Create a matrix S with its columns created from the terms in the sine series:

$$\sin x = x - \frac{x^3}{3!} + \frac{x^5}{5!} - \frac{x^7}{7!} + \cdots$$

. First create a column vector x between 0 and 2π with the command
`x = linspace(0,2*pi,101)'`. Now compute
`A = [x  -x.^3/6  x.^5/120  -x.^7/factorial(7)];`

   - Save matrix A and vector x in a Mat-file.
   - Clear all variables from the workspace.
   - Select the Mat-file in the **Current Directory** pane and then double-click on the variable x visible in the file contents pane below the directory pane to load x in the workspace. This action is equivalent to typing `load` *Mat-file* x.
   - Double-click on the variable x visible in the workspace pane to open the **Variable Editor** and inspect the entries of x. You can edit any value if you wish.
   - Plot $\sin(x)$ with the command `plot(x,sin(x)), grid, hold on`.
   - Now load variable `A` from the Mat-file and plot `A(:,1)`, `A(:,1)+A(:,2)`, `A(:,1)+A(:,2)+A(3,:)`, and `sum(A,2)` on the existing graph. Adjust the axes with the command `axis([0 2*pi -2 2]`. You can compare the difference in $\sin(x)$ and its series representation with increasing terms (up to the fourth term) on this graph.
   - Save this plot as a JPEG. You can do this with the command `print -djpeg sineseriesplot.jpeg` or by clicking on **File → Save As...** and choosing JPEG as the file format. (We will use this image in the next exercise.)

2. **Loading image data from graphic format files:** You can import image data, written using various file formats, into MATLAB fairly easily. There are several functions provided in MATLAB to import image, audio, and video files. Here, we will import data from a JPEG file and reconstruct the image from the data.

   - First, learn a little more about importing data by typing `help importdata`.
   - Now import the image data in an array S from the file sineseriesplot.jpeg you created earlier. Check the size of S to verify that it is a 3-D matrix (corresponding to three colors—red, green, and blue).
   - Reconstruct the image with the command `image(S)`.
   - Try reading some other JPEG file that you may have (a picture perhaps), store the data in some variable, and reconstruct the picture with the `image` command. You have to play with setting the axes correctly to get the image right.

2.10 Lesson 10: Working with Files and Directories

Goal: To learn how to navigate through MATLAB directory system and work with files in a particular location.

Time Estimates
Lesson: 15 minutes
Exercises: 20 minutes

What you are going to learn

- How to find your bearings in the jungle of directories.
- How to see the contents of a directory.
- How to create a new directory and change the current directory.
- How to copy a file from one directory to another.
- How to see the list of only MATLAB-related files.

Method: MATLAB includes several menu-driven features that make file navigation much easier (compared with the earlier versions). You will explore some of these features in the exercises. In the lesson, you will learn commands that pretty much do the same thing from the command line. The commands that you will use are `pwd, dir, ls, cd, what, makedir`, etc. Please go ahead and try the commands shown in Fig. 2.14.

Comments:

- It is important to know which directory you are currently in because any files you write or save will, by default, be saved into the current directory.
- You may have your M-files—scripts or functions—stored in a different directory than the current directory. In that case, MATLAB will not be able to access your files by itself. You will have to either change the current directory to the directory where your files are or include your directory in the MATLAB *path*. The `path` is a MATLAB variable that contains the full path names of all directories that MATLAB can access during a work session. It is created during MATLAB installation. You can, however, modify it. See on-line help on `path`.
- You can create new directories from within the MATLAB command window. You can also copy or move files from one directory to another with MATLAB commands (`copyfile, movefile`). See on-line help on `pwd, cd, mkdir, rmdir, copyfile, movefile, what`, and `which`.

```
>> pwd                                    Print (show) working directory.

ans =

C:\matlab\work

>> dir                                    Show the contents of the directory.

   .    ..     circle.m  circlefn.m

>> mkdir lesson10dir                      Make a new directory named
>> ls                                     lesson10dir.  List the contents of
                                          the directory again.

circle.m  circlefn.m       lesson10dir

>> cd lesson10dir                         Change directory to lesson10dir
>> ls                                     and list its contents.

>> x = 1:100;                             Create some data,  x and y, and
>> y = log10(x);                          save them in a datafile named
>> save xydatafile x y                    xydatafile.  List the contents
>> ls                                     of the directory again.

xydatafile.mat

>> what                                   List MATLAB-related files.

MAT-files in the current directory C:\matlab\work\le...

xydatafile

>> cd ..                                  Change directory one level up.
>> pwd                                    Note the space between cd and ..
                                          Check the working directory name.
ans =

C:\matlab\work

>> copyfile circle.m lesson10dir\
>> ls lesson10dir
                                          Copy the file circle.m into the
                                          directory lesson10dir.
circle.m  xydatafile.mat                  List the contents of lesson10dir.
```

Figure 2.14: Lesson 10: Working with files and directories.

EXERCISES

1. **Working with directories:** Information about the current directory's name and its contents are available in the MATLAB desktop. Look at the command window toolbar. There is a small window that shows the name of the current directory, along with its path. For example, Fig. 2.15 shows that the current directory is /Users/rudrapra/Documents/MATLAB (this is on a Mac; on a PC it may be C:\Documents and Settings\...\My Dcuments\MATLAB). As the path indicates, it is inside the MATLAB directory.

   The current directory is also displayed in a separate subwindow to the left of the command window. If it is not visible, click on the Current Directory tab. This subwindow also lists the contents of the current directory.

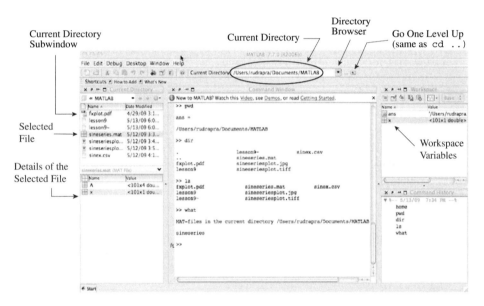

Figure 2.15: Directory information in the MATLAB desktop. The current directory is always displayed in the command window toolbar. In addition, the directory and its contents are displayed in the current directory subwindow.

   In this exercise, the idea is to use the GUI (graphical user interface) buttons in the MATLAB desktop to change directories, see their contents, create directories, etc. These are the tasks that you did using commands in the tutorial lesson (Fig. 2.14).

   - Go to the current directory subwindow and change the directory using the One Level Up or the Directory Browser button. Try several directories.
   - Create a new directory using the New Directory button.
   - Move or copy a file from one directory to another.

   Now you know your way around the directories using mouse clicks.

2.11   Lesson 11: Publishing Reports

Goal: To learn how to use MATLAB's built-in *publisher* for publishing reports of
your MATLAB work as attractive HTML or MS Word documents (other options
such as LaTeX and PowerPoint are also available).

Time Estimates
Lesson: 20 minutes
Exercises: 30 minutes

What you are going to learn

- How to create a cell script for MATLAB publisher.
- How to execute the script.
- How to publish the script and its results in an HTML file.
- How to publish the script and its results in a Word file.
- How to learn more about fancy formatting for the publisher.

Method: MATLAB includes an automatic report generator called *publisher*.
This publisher is accessible, like many other utilities, both from the menu and
the command line. The publisher *publishes a script* in several formats, including
HTML, XML, MS Word, PowerPoint, and LaTeX. Here you will write a simple
script and publish it both as an HTML document and a Word document. Follow
the steps below.

1. First, open a new file in the editor and enter the following lines (including
   blank lines). The only thing new here is the double percent (%%) character.
   It indicates the beginning of a new *cell*, a unit of a set of commands (see
   Exercises). The text in a line beginning with %% is used as the title of that
   section (or cell) by the publisher.

   ```
   %% Publishing Reports - A Simple Example
   %% A Spiral Plot
   % Let us plot a spiral given by
   % r(t) = exp(-theta/10), 0<=theta<=10*pi

   %% Create vectors theta and r
   theta = linspace(0,10*pi,200); % 200 linearly spaced points between
                                  %- 0 and 10*pi
       r = exp(-theta/10);        % compute r

   %% Now plot theta vs r using polar plot
   polar(theta,r)
   ```

2. Save this script file as, say, spiralplot.m in the current directory.

3. Now use the following command to publish your script, including the resulting
   plot, into an HTML file and open the published file.

```
>> publish('spiralplot','html')        Publish to HTML format.

>> cd html; open spiralplot.html       Open the published file.
```

4. Voilà! You should see a nice, colorful HTML file like the one shown in Fig. 2.16.

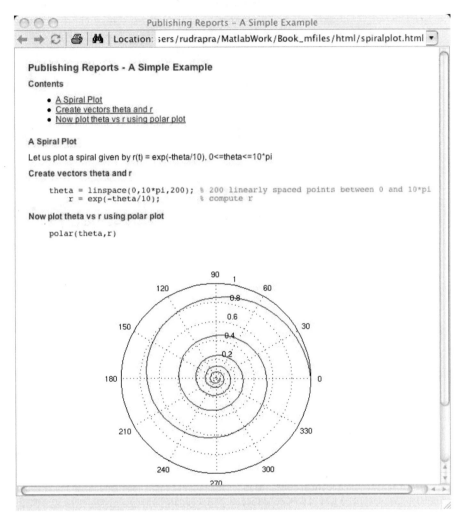

Figure 2.16: An HTML file created by the MATLAB publisher from the script file spiralplot.m shown on the previous page.

5. Now see the on-line help on publish and use this command to publish a Word document (if you are on a PC).

EXERCISES

1. **Using the publisher from the editor window:** Open the file spiralplot.m that you created while working with Lesson 11. You can open the file by following File → Open from the menu or from the command window with either `edit spiralplot.m` or `open spiralplot.m`. Once the file is open in the editor, select Publish from the File menu (see Fig. 2.17) of the editor window, or click on the publish icon (next to the printer icon in the menu bar). The script open in the editor will be published to an HTML file and MATLAB will automatically open the HTML file for your perusal.

Figure 2.17: The MATLAB publisher can be invoked from the File menu when the script is open in the editor window.

2. **What are those cells?** A cell, in a script file, is a set of commands enclosed between two cell separators (lines that start with %%). You can execute a cell independent of other cells. You can also execute one cell and then the next by simply clicking on the cell execution option buttons (see Fig. 2.17) in the toolbar of the editor. Move your mouse on these buttons to see what each one does.

    Now, open the spiralplot.m file again (if you have closed it) in the editor and add the following two cells at the end of the file. You can just type %% to start a new cell or click on the Insert Cell Divider button in the toolbar to insert it.

```
%% Polar to Cartesian coordinates
    x = r.*cos(theta);
```

```
    y = r.*sin(theta);
    plot(x,y), axis('equal')
%% Polar to Cartesian with pol2cart
    [xx,yy] = pol2cart(theta,r);
    plot(xx,yy), axis('equal')
```

Now experiment with executing one cell at a time by clicking on the appropriate cell execution button in the toolbar. If you are happy, publish the script again. The newly published document should have three plots, one polar and two Cartesian, showing the same spiral.

3. Just a bit of fun with formatting the script: It is possible to format your report to make it look better. In particular, you can introduce bold text, monospaced text, bulleted list, and nicely formatted mathematical equations. Out of these, formatting equations is something that definitely improves the readability of your report. However, it also requires learning a little bit of LaTeX.[1] See Section 4.5 on page 129 for more discussion.

Let us introduce a bit of formatting in the script file so that it shows equations for the spiral and polar to Cartesian conversion formulas as mathematical equations. Edit **spiralplot.m** so that it now looks like the following (note, there are basically two changes for the mathematical equations).

```
%% Publishing Reports - A Simple Example
%% A Spiral Plot
% Let us plot a spiral given by
%%
% $$r(\theta) = e^{-\frac{\theta}{10}}, \quad 0\le\theta\le 10\pi$$
%% Create vectors theta and r
    .

    .       (leave the other lines unchanged)
%% Polar to Cartesian coordinates
% using the usual formulas
%%
% $$ x = r\cos\theta, \quad y = r \sin\theta.$$
%
    x = r.*cos(theta);
    y = r.*sin(theta);
    plot(x,y), axis('equal')
```

Now save the file and publish it again. Look at the output. It should have nice formatted equations that look like

$$r(\theta) = e^{-\frac{\theta}{10}}, \quad 0 \le \theta \le 10\pi$$

and

$$x = r\cos\theta, \quad y = r\sin\theta.$$

4. Invoke `helpdesk`, do a search for *publish*, and read (and watch videos) as much as you would like to learn about the publisher.

---

[1]If you have not heard of LaTeX, it's time to Google. For a minimal list of LaTeX commands, see Table 4.1 on page 131.

𝟑. *Interactive Computation*

In principle, one can do all calculations in MATLAB interactively by entering commands sequentially in the command window, although a *script file* (explained in Section 4.1) is perhaps a better choice for computations that involve more than a few steps. The interactive mode of computation, however, makes MATLAB a powerful scientific calculator that puts hundreds of built-in mathematical functions for numerical calculations and sophisticated graphics at the fingertips of the user.

In this chapter, we introduce you to some of MATLAB's built-in functions and capabilities, through examples of interactive computation. The basic things to keep in mind are:

Where to type commands: All MATLAB commands or expressions are entered in the command window at the MATLAB prompt (≫).

How to execute commands: To execute a command or statement, you must press return or enter at the end.

What to do if the command is very long: If your command does not fit on one line, you can continue the command on the next line if you type three consecutive periods at the end of the first line. You can keep continuing this way until the length of your command hits the limit, which is 4,096 characters. For more information, see the discussion on continuation on pages 65 and 112.

How to name variables: Names of variables must begin with a letter. After the first letter, any number of digits or underscores may be used, but MATLAB remembers only the first 31 characters.

What is the precision of computation: All computations are carried out internally in double precision unless specified otherwise. The appearance of numbers on the screen, however, depends on the `format` in use (see next item).

How to control the display format of the output: The output appearance of floating-point numbers (number of digits after the decimal, etc.) is controlled

format of the output.

with the `format` command. The default is `format short`, which displays four digits after the decimal. For other available formats and how to change them, see Section 1.6.3, page 7, or on-line help on `format`.

How to suppress the screen output: A semicolon (;) at the end of a command suppresses the screen output, although the command is carried out and the result is saved in the variable assigned to the command or in the default variable `ans`.

How to recall previously typed commands: Use the up-arrow key to recall previously typed commands. MATLAB uses *smart recall*, so you can also type one or two letters of your command and use the up-arrow key for recalling the command starting with those letters. Also, all your commands are recorded in the command history subwindow (see Fig. 1.3 on page 8). You can double-click on any command in the command history subwindow to execute it in the command window.

How to set paged-screen display: For paged-screen display (one screenful of output display at a time) use the command `more on`.

Where and how to save results: If you need to save some of the computed results for later processing, you can save the variables in a file in binary or ASCII format with the `save` command. See Section 3.7, page 90, for more information.

How to save figures: You can save a figure in a .fig file by selecting File → Save As from the figure window. Once you save it, you can open it later in MAT-LAB using `open` command or by selecting File → Open... from the MATLAB main menu. You can also save figures in one of numerous formats for printing or exporting to other applications (see Section 6.6, page 219).

How to print your work: You can print your entire session in MATLAB, part of it, or selected segments of it, in one of several ways. The simplest way, perhaps, is to create a diary with the `diary` command (see Section 3.7.3 for more information) and save your entire session in it. Then you can print the diary just the way you would print any other file on your computer. On PCs and Macs, however, you can print the session by selecting Print from the File menu. (Before you print, make sure that the command window is the active window. If it isn't, just click on the command window to make it active).

What about comments: MATLAB takes anything following a % as a comment and ignores it.[1] You are not likely to use a lot of comments while computing interactively, but you will use them when you write programs in MATLAB.

Because MATLAB derives most of its power from matrix computations and assumes every variable to be, at least potentially, a matrix, we start with descriptions and examples of how to enter, index, manipulate, and perform some useful calculations with matrices.

---

[1]The exception to this rule is when the % appears in a quote-enclosed character string or in certain input/output format statements.

3.1   Matrices and Vectors

3.1.1   Input

For on-line help type:
`help elmat`

A matrix is entered row-wise, with consecutive elements of a row separated by a space or a comma, and the rows separated by semicolons or carriage returns. The entire matrix must be enclosed within square brackets. Elements of the matrix may be real numbers, complex numbers, or valid MATLAB expressions.[2]

Examples:

Matrix	**MATLAB input command**
$A = \begin{bmatrix} 1 & 2 & 5 \\ 3 & 9 & 0 \end{bmatrix}$	`A = [1 2 5;  3 9 0]`
$B = \begin{bmatrix} 2x & \ln x + \sin y \\ 5i & 3 + 2i \end{bmatrix}$	`B = [2*x log(x)+sin(y); 5i 3+2i]` (see the note in the margin.)

For the matrix B command to work, variables x and y must be predefined, e.g., `x=1; y=2;`, etc.

Special cases: Vectors and scalars

- A vector is a special case of a matrix, with just one row or one column. It is entered the same way as a matrix.
  Examples: `u = [1 3 9]` produces a row vector, and
  `v = [1; 3; 9]` produces a column vector.
- A scalar does not need brackets.
  Example: `g = 9.81;`
- Square brackets with no elements between them create a null matrix.
  Example: `X = []`. (See Fig. 3.1 for a more useful example).

Continuation

If it is not possible to type the entire input on the same line, then use three consecutive periods (...) to signal continuation and continue the input on the next line. The three periods are called an *ellipsis*. For example,

```
A = [1/3    5.55*sin(x)   9.35   0.097;...
     3/(x+2*log(x))    3    0    6.555; ...
     (5*x-23)/55    x-3    x*sin(x)    sqrt(3)];
```

produces the intended 3×4 matrix A (provided, of course, x has been assigned a value before). A matrix can also be entered across multiple lines using carriage returns at the end of each row. In this case, the semicolons and ellipses at the end of each row may be omitted. Thus, the following three commands are equivalent:

```
A = [1 3 9; 5 10 15; 0 0 -5];
A = [1 3 9
     5 10 15
     0 0 -5];
A = [1 3 9; 5 10 ...
     15; 0 0 -5];
```

---

[2]The box on the right lets you know that you can learn more about this topic from MATLAB's on-line help, on the help topic called *elmat* (for *el*ementary *mat*rix manipulations).

Continuation across several input lines achieved through an ellipsis is not limited to matrix input. This construct may be used for other commands and for a long list of command arguments (see Section 4.3.2, page 112), as long as the command does not exceed 4,096 characters.

3.1.2 Indexing (or subscripting)

Once a matrix exists, its elements are accessed by specifying their row and column indices. Thus, `A(i,j)` in MATLAB refers to the element a_{ij} of matrix A, i.e., the element in the ith row and jth column. This notation is fairly common in computational software packages and programming languages. MATLAB, however, provides a much higher level of index specification—it allows a range of rows and columns to be specified at the same time. For example, the statement `A(m:n,k:l)` specifies rows m to n and columns k to l of matrix A. When the rows (or columns) to be specified range over all rows (or columns) of the matrix, a colon can be used as the row (or column) index. Thus, `A(:,5:20)` refers to the elements in columns 5 through 20 of *all* the rows of matrix A. This feature makes matrix manipulation much easier and provides a way to take advantage of the "vectorized" nature of calculations in MATLAB. (See Fig. 3.1 for examples.)

Dimension

Matrix dimensions are determined automatically by MATLAB, i.e., no explicit dimension declarations are required. The dimensions of an existing matrix A may be obtained with the command `size(A)` or more explicitly with `[m,n]=size(A)`, which assigns the number of rows and columns of A to the variables m and n, respectively. When a matrix is entered by specifying a single element or a few elements of the matrix, MATLAB creates a matrix just big enough to accommodate the elements. Thus, if the matrices B and C do not exist already, then

$$\texttt{B(2,3) = 5;} \qquad \text{produces} \qquad B = \begin{bmatrix} 0 & 0 & 0 \\ 0 & 0 & 5 \end{bmatrix},$$

$$\texttt{C(3,1:3) = [1 2 3];} \quad \text{produces} \quad C = \begin{bmatrix} 0 & 0 & 0 \\ 0 & 0 & 0 \\ 1 & 2 & 3 \end{bmatrix}.$$

3.1.3 Matrix manipulation

As you can see from examples in Fig. 3.1, it is fairly easy to correct wrong entries of a matrix, extract any part or submatrix of a matrix, or delete or add rows and columns. These manipulations are done with MATLAB's smart indexing feature. By specifying vectors as the row and column indices of a matrix, one can reference and modify any submatrix of a matrix. Thus, if A is a 10×10 matrix, B is a 5×10 matrix, and y is a 20 elements long row vector, then

$$\texttt{A([1 3 6 9],:) = [B(1:3,:); y(1:10)]}$$

```
>> A = rand(4,3)                          Create a 4 x 3 random matrix A.

A =
     0.8147      0.6324      0.9575
     0.9058      0.0975      0.9649
     0.1270      0.2785      0.1576
     0.9134      0.5469      0.9706

>> A(3:4, 2:3)                            Get those elements of A that are
                                          located in rows 3 to 4 and columns
ans =                                     2 to 3.
     0.2785      0.1576
     0.5469      0.9706

>> A(:,4) = A(:,1)                        Add a fourth column to A and set
                                          it equal to the first column of A.
A =
     0.8147      0.6324      0.9575      0.8147
     0.9058      0.0975      0.9649      0.9058
     0.1270      0.2785      0.1576      0.1270
     0.9134      0.5469      0.9706      0.9134

                                          Replace the last 3 x 3 submatrix
>> A(2:4,2:4) = eye(3)                     of A (rows 2 to 4,  columns 2 to 4)
                                          by a 3 x 3 identity matrix.
A =
     0.8147      0.6324      0.9575      0.8147
     0.9058      1.0000           0           0
     0.1270           0      1.0000           0
     0.9134           0           0      1.0000

>> A([1 3],:) = []                        Delete the first and third rows of A.

A =
     0.9058      1.0000           0           0
     0.9134           0           0      1.0000

>> A = round(A)                           Round off all entries of A.

A =
        1        1        0        0
        1        0        0        1

>> A(:)'                                  String out all elements of A in a
                                          row (note the transpose at the end).
ans =
        5        1        1        0        0        1
```

Figure 3.1: Examples of matrix input and matrix index manipulation.

replaces first, third, and sixth rows of A by the first three rows of B, and the ninth row of A by the first 10 elements of y. In such manipulations, it is imperative that the sizes of the submatrices to be manipulated are compatible. For example, in the previous assignment, the number of columns in A and B must be the same, and the total number of rows specified on the right-hand side must be the same as the number of rows specified on the left.

Thus, from a given matrix, picking out either a range of rows or columns or a set of noncontiguous rows or columns is straightforward. This can be done by creating index vectors with numbers representing the desired rows or columns.

$$\text{So, if} \quad Q = \begin{bmatrix} 2 & 3 & 6 & 0 & 5 \\ 0 & 0 & 20 & -4 & 3 \\ 1 & 2 & 3 & 9 & 8 \\ 2 & -5 & 5 & -5 & 6 \\ 5 & 10 & 15 & 20 & 25 \end{bmatrix} \quad \text{and} \quad v = \begin{bmatrix} 1 & 4 & 5 \end{bmatrix},$$

$$\text{then} \quad \text{Q(v,:)} = \begin{bmatrix} 2 & 3 & 6 & 0 & 5 \\ 2 & -5 & 5 & -5 & 6 \\ 5 & 10 & 15 & 20 & 25 \end{bmatrix} \quad \text{and} \quad \text{Q(:,v)} = \begin{bmatrix} 2 & 0 & 5 \\ 0 & -4 & 3 \\ 1 & 9 & 8 \\ 2 & -5 & 6 \\ 5 & 20 & 25 \end{bmatrix}.$$

Logicals (0-1) in the matrix index: Only nonzero positive integers are legal index entries for a matrix. The only exception is the *logical* 0 (zero). You can use vectors made up of zeros and ones in the matrix index if

1. The vector is produced by logical or relational operations (see Sections 3.2.2 and 3.2.3).

2. The 0-1 vector created by you is converted into a logical array with the command `logical`.

In either case, the index vector picks out rows or columns corresponding to the location of ones. For example, to get the first, fourth, and fifth rows of Q with 0-1 vectors, you can do

```
v = [1 0 0 1 1];      v = logical(v);      Q(v,:).
```

Reshaping matrices

Matrices can be reshaped into a vector or any other appropriately sized matrix:

As a vector: All the elements of matrix A can be strung into a single-column vector b by the command `b = A(:)` (matrix A is stacked in vector b columnwise).

As a differently sized matrix: If matrix A is an $m \times n$ matrix, it can be reshaped into a $p \times q$ matrix, as long as $m \times n = p \times q$, with the command `reshape(A,p,q)`. Thus, for a 6×6 matrix A,

`reshape(A,9,4)`	transforms A into a 9×4 matrix, and
`reshape(A,3,12)`	transforms A into a 3×12 matrix.

Now let us look at some frequently used manipulations.

Transpose

The transpose of matrix A is obtained by typing `A'`, i.e., the name of the matrix followed by the single right quote. For a real matrix A, the command `B=A'` produces $B = A^T$, that is, $b_{ij} = a_{ji}$, and for a complex matrix A, `B=A'` produces the conjugate transpose $B = \bar{A}^T$, that is, $b_{ij} = \bar{a}_{ji}$.

Examples:

If $A = \begin{bmatrix} 2 & 3 \\ 6 & 7 \end{bmatrix}$, then `B = A'` gives $B = \begin{bmatrix} 2 & 6 \\ 3 & 7 \end{bmatrix}$.

If $C = \begin{bmatrix} 2 & 3+1i \\ 6i & 7i \end{bmatrix}$, then `Ct = C'` gives $Ct = \begin{bmatrix} 2 & -6i \\ 3-1i & -7i \end{bmatrix}$.

If $u = [0 \ 1 \ 2 \ \cdots \ 9]$, then `v = u(3:6)'` gives $v = \begin{bmatrix} 2 \\ 3 \\ 4 \\ 5 \end{bmatrix}$.

Initialization

Initialization of a matrix is not necessary in MATLAB. However, it is advisable in the following two cases.

1. **Large matrices:** If you are going to generate or manipulate a large matrix, initialize the matrix to a zero matrix of the required dimension. An $m \times n$ matrix can be initialized by the command `A=zeros(m,n)`. The initialization reserves for the matrix a contiguous block in the computer's memory. Matrix operations performed on such matrices are generally more efficient.

2. **Dynamic matrices:** If the rows or columns of a matrix are computed in a loop (e.g., `for` or `while` loop) and appended to the matrix (see the following) in each execution of the loop, then you might want to initialize the matrix to a null matrix before the loop starts. A null matrix A is created by the command `A=[]`. Once created, a row or column of any size may be appended to A as described next.

Appending a row or column

A row can be easily appended to an existing matrix, provided the row has the same length as the length of the rows of the existing matrix. The same thing goes for columns. The command `A=[A; u]` appends a row vector u to the rows of A, while `A=[A v]` appends a column vector v to the columns of A. A row or column of any size may be appended to a null matrix.

Examples: If

$$A = \begin{bmatrix} 1 & 0 & 0 \\ 0 & 1 & 0 \\ 0 & 0 & 1 \end{bmatrix}, \quad u = \begin{bmatrix} 5 & 6 & 7 \end{bmatrix}, \text{ and } v = \begin{bmatrix} 2 \\ 3 \\ 4 \end{bmatrix},$$

then

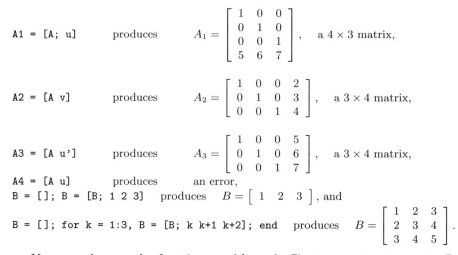

A1 = [A; u]   produces $A_1 = \begin{bmatrix} 1 & 0 & 0 \\ 0 & 1 & 0 \\ 0 & 0 & 1 \\ 5 & 6 & 7 \end{bmatrix}$,   a 4×3 matrix,

A2 = [A v]   produces $A_2 = \begin{bmatrix} 1 & 0 & 0 & 2 \\ 0 & 1 & 0 & 3 \\ 0 & 0 & 1 & 4 \end{bmatrix}$,   a 3×4 matrix,

A3 = [A u']   produces $A_3 = \begin{bmatrix} 1 & 0 & 0 & 5 \\ 0 & 1 & 0 & 6 \\ 0 & 0 & 1 & 7 \end{bmatrix}$,   a 3×4 matrix,

A4 = [A u]   produces   an error,

B = []; B = [B; 1 2 3]   produces $B = \begin{bmatrix} 1 & 2 & 3 \end{bmatrix}$, and

B = []; for k = 1:3, B = [B; k k+1 k+2]; end   produces $B = \begin{bmatrix} 1 & 2 & 3 \\ 2 & 3 & 4 \\ 3 & 4 & 5 \end{bmatrix}$.

You can also use the function cat(dim, A, B) to concatenate matrix B to matrix A along the dimension dim (for 2-D matrices, $dim=1$ for rows and $dim=2$ for columns). Of course, the dimensions of the two matrices must be compatible.

Deleting a row or column

Any row(s) or column(s) of a matrix can be deleted by setting the row or column to a null vector.

Examples:

```
A = rand(6); u = rand(10,1);    create a 6 × 6 matrix A and 10 × 1 vector u,
A(2,:)   = []                   deletes the second row of matrix, A,
A(:,3:5) = []                   deletes the third through fifth columns of A,
A([1 3],:) = []                 deletes the first and the third row of A, and
u(5:length(u)) = []             deletes all elements of vector u except 1 through 4.
```

For on-line help type:
`help elmat`

Utility matrices

To aid matrix generation and manipulation, MATLAB provides many useful utility matrices. For example,

eye(m,n)	returns an m by n matrix with ones on the main diagonal,
zeros(m,n)	returns an m by n matrix of zeros,
ones(m,n)	returns an m by n matrix of ones,
rand(m,n)	returns an m by n matrix of random numbers,
randn(m,n)	returns an m by n matrix of normally distributed numbers,
diag(v)	generates a diagonal matrix with vector v on the diagonal,
diag(A)	extracts the diagonal of matrix A as a vector, and
diag(A,1)	extracts the first upper off-diagonal vector of matrix A.

The first four commands with a single argument, e.g., ones(m), produce square matrices of dimension m. For example, eye(3) produces a 3×3 identity matrix. A matrix can be built with many block matrices as well. See examples in Fig. 3.2.

Here is a list of some more functions used in matrix manipulation:

```
>> eye(3)

ans =

     1     0     0
     0     1     0
     0     0     1
```

eye(n) creates an $n \times n$ identity matrix. The commands zeros, ones, and rand work in a similar way.

```
>> B = [ones(3) zeros(3,2); zeros(2,3) 4*eye(2)]

B =

     1     1     1     0     0
     1     1     1     0     0
     1     1     1     0     0
     0     0     0     4     0
     0     0     0     0     4
```

Create a matrix B using submatrices made up of elementary matrices: ones, zeros, and the identity matrix of the specified sizes.

```
>> diag(B)'

ans =

     1     1     1     4     4
```

This command pulls out the diagonal of B in a row vector. Without the transpose, the result would obviously be a column vector.

```
>> diag(B,1)'

ans =

     1     1     0     0
```

The second argument of the command specifies the off-diagonal vector to be pulled out. Here we get the first upper off-diagonal vector. A negative value of the argument specifies the lower off-diagonal vectors.

```
>> d = [2 4 6 8];
>> d1 = [-3 -3 -3];
>> d2 = [-1 -1];
>> D = diag(d) + diag(d1,1) + diag(d2,-2)
```

Create vectors d, $d1$, and $d2$ of length 4, 3, and 2, respectively.

```
D =

     2    -3     0     0
     0     4    -3     0
    -1     0     6    -3
     0    -1     0     8
```

Create a matrix D by putting d on the main diagonal, $d1$ on the first upper diagonal, and $d2$ on the second lower diagonal.

Figure 3.2: Examples of matrix manipulation using utility matrices and functions.

`rot90`	rotates a matrix by 90°,
`fliplr`	flips a matrix from left to right,
`flipud`	flips a matrix from up to down,
`tril`	extracts the lower triangular part of a matrix,
`triu`	extracts the upper triangular part of a matrix, and
`reshape`	changes the shape of a matrix.

Special matrices

There is also a set of built-in special matrices such as `hadamard, hankel, hilb, invhilb, kron, pascal, toeplitz, vander, magic`, etc. For a complete list and help on these matrices, type `help elmat`. (In earlier versions of MATLAB, these matrices were under `specmat`.)

3.1.4   Creating vectors

Very often we need to create a vector of numbers over a given range with a specified increment. The general command to do this in MATLAB is

$$\boxed{\texttt{v = }\mathit{InitialValue}\texttt{:}\mathit{Increment}\texttt{:}\mathit{FinalValue}}$$

The three values in the preceding assignment can also be valid MATLAB expressions. If no increment is specified, MATLAB uses the default increment of 1.

Examples:

`a = 0:10:100`	produces $a = [\ 0 \quad 10 \quad 20 \quad \ldots \quad 100\ ]$,
`b = 0:pi/50:2*pi`	produces $b = [\ 0 \quad \frac{\pi}{50} \quad \frac{2\pi}{50} \quad \ldots \quad 2\pi\ ]$, i.e., a linearly spaced vector from 0 to 2π spaced at $\pi/50$,
`u = 2:10`	produces $a = [\ 2 \quad 3 \quad 4 \quad \ldots \quad 10\ ]$.

As you may notice, no square brackets are required if a vector is generated this way; however, a vector assignment such as `u=[1:10 33:-2:19]` does require square brackets to force the concatenation of the two vectors: `[1 2 3 ... 10]` and `[33 31 29 ... 19]`. Finally, we mention the use of two frequently used built-in functions to generate vectors:

→ `linspace(a,b,n)` generates a linearly spaced vector of length n from a to b.
   Example: `u=linspace(0,20,5)` generates `u=[0 5 10 15 20]`.
   Thus, `u=linspace(a,b,n)` is the same as `u=a:(b-a)/(n-1):b.`

`logspace(a,b,n)` generates a logarithmically spaced vector of length n from 10^a to 10^b.
   Example: `v=logspace(0,3,4)` generates `v=[1 10 100 1000]`.
   Thus, `logspace(a,b,n)` is the same as `10.^(linspace(a,b,n))`. (The array operation `.^` is discussed in the next section.)

Special vectors, such as vectors of zeros or ones of a specific length, can be created with the utility matrix functions `zeros, ones`, etc.

Examples:

`u = zeros(1,1000)`	initializes a 1000-element-long row vector, and
`v = ones(10,1)`	creates a 10-element-long column vector of ones.

3.2 Matrix and Array Operations

3.2.1 Arithmetic operations

For people who are used to programming in a conventional language such as Pascal, Fortran, or C, it is an absolute delight to be able to write a matrix product as `C=A*B` where A is an $m \times n$ matrix and B is an $n \times k$ matrix. MATLAB allows

For on-line help type:
`help ops`

+	addition
−	subtraction
∗	multiplication
/	division
^ (caret)	exponentiation

For on-line help type:
`help arith`

to be carried out on matrices in straightforward ways as long as the operation makes sense mathematically and the operands are compatible. Thus,

`A+B` or `A-B`   is valid if A and B are of the same size,

`A*B`   is valid if A's number of columns equals B's number of rows,

`A/B`   is valid and equals $A \cdot B^{-1}$ for same-size square matrices, and

`A^2`   which equals `A*A`, makes sense only if A is square.

In all these commands, if B is replaced by a scalar, say α, the arithmetic operations are still carried out. In this case, the command `A+`α adds α to each element of A, the command `A*`α (or α`*A`) multiplies each element of A by α and so on. Vectors are just treated as a single row or a column matrix and therefore a command such as `w=u*v`, where u and v are same-size vectors, say $m \times 1$, produces an error (because you cannot multiply an $m \times 1$ matrix with an $m \times 1$ matrix), whereas `w=u*v`$'$ and `w=u`$'$`*v` execute correctly, producing the outer and inner products of the two vectors, respectively (see examples in Fig 2.7 on page 41).

The left division: In addition to the normal or *right* division (/), there is a *left* division (\\) in MATLAB. This division is used to solve a matrix equation. In particular, the command `x=A\b` solves the matrix equation $Ax = b$. Thus `A\b` is *almost* the same as `inv(A)*b` but faster and more numerically stable than computing `inv(A)*b`. In the degenerate case of scalars, `5\3` gives 0.6, which is 3/5 or $5^{-1} * 3$.

Array operation: How does one get products such as $[u_1 v_1 \quad u_2 v_2 \quad u_3 v_3 \quad \dots \quad u_n v_n]$ from two vectors u and v? No, you do not have to use `for` loops. You can do *array operation*—operations done on an element-by-element basis. Element-by-element multiplication, division, and exponentiation between two matrices or vectors of the same size are done by preceding the corresponding arithmetic operators by a period:

For on-line help type:
`help arith`
`help slash`

.∗	element-by-element multiplication
./	element-by-element left division
.\\	element-by-element right division
.^	element-by-element exponentiation
.'	nonconjugated transpose

Examples: u.*v   produces $[u_1v_1 \ u_2v_2 \ u_3v_3 \ \ldots],$

u./v   produces $[u_1/v_1 \ u_2/v_2 \ u_3/v_3 \ \ldots],$ and

u.^v   produces $[u_1^{v_1}, \ u_2^{v_2}, \ u_3^{v_3}, \ \ldots].$

The same is true for matrices. For two same-sized matrices A and B, the command C=A.*B produces a matrix C with elements $C_{ij} = A_{ij}B_{ij}$. Clearly, there is a big difference between A^2 and A.^2 (see Fig. 2.7 on page 41). Once again, scalars do enjoy a special status. For example, 1./v happily computes $[1/v_1 \ 1/v_2 \ 1/v_3 \ \ldots]$, and pi.^v gives $[\pi^{v_1} \ \pi^{v_2} \ \pi^{v_3} \ \ldots]$, whereas u./v or u.^v produces an error if u and v are not of the same size. See Section 3.4 on page 81 for more discussion on array operations.

For on-line help type:

help relop

3.2.2   Relational operations

There are six relational operators in MATLAB:

<	less than
<=	less than or equal
>	greater than
>=	greater than or equal
==	equal
~=	not equal

These operations result in a vector or matrix of the same size as the operands, with 1 where the relation is true and 0 where it is false.

Examples: If $x = [1 \ 5 \ 3 \ 7]$ and $y = [0 \ 2 \ 8 \ 7]$, then

k = x < y	results in $k = [0 \ 0 \ 1 \ 0]$	because $x_i < y_i$ for $i = 3$,
k = x <= y	results in $k = [0 \ 0 \ 1 \ 1]$	because $x_i \le y_i$ for $i = 3$ and 4,
k = x > y	results in $k = [1 \ 1 \ 0 \ 0]$	because $x_i > y_i$ for $i = 1$ and 2,
k = x >= y	results in $k = [1 \ 1 \ 0 \ 1]$	because $x_i \ge y_i$ for $i = 1, 2,$ and 4,
k = x == y	results in $k = [0 \ 0 \ 0 \ 1]$	because $x_i = y_i$ for $i = 4$, and
k = x ~= y	results in $k = [1 \ 1 \ 1 \ 0]$	because $x_i \ne y_i$ for $i = 1, 2,$ and 3.

Although these operations are usually used in conditional statements such as *if-then-else* to branch out to different cases, they can be used to do pretty sophisticated matrix manipulation. For example, u=v(v>=sin(pi/3)) finds all elements of vector v such that $v_i \ge \sin\frac{\pi}{3}$ and stores them in vector u. Two or more of these operations can also be combined with the help of *logical operators* (described next).

For on-line help type:

help relop

3.2.3   Logical operations

There are four logical operators:

&	logical AND
|	logical OR
~	logical complement (NOT)
xor	exclusive OR

These operators work in a similar way as the relational operators and produce vectors or matrices of the same size as the operands, with 1 where the condition is true and 0 where false.

Examples: For two vectors $x = [0\ \ 5\ \ 3\ \ 7]$ and $y = [0\ \ 2\ \ 8\ \ 7]$,

m = (x>y)&(x>4)	results in $m = [0\ 1\ 0\ 0]$, because the condition is true only for x_2,
n = x|y	results in $n = [0\ 1\ 1\ 1]$, because either x_i or y_i is nonzero for $i = [2\ 3\ 4]$,
m = ~(x|y)	results in $m = [1\ 0\ 0\ 0]$, which is the logical complement of x|y, and
p = xor(x,y)	results in $p = [0\ 0\ 0\ 0]$, because there is no such index i
	for which x_i or y_i, but not both, is nonzero.

Because the output of the logical operations is a 0-1 vector or a 0-1 matrix, the output can be used as the index of a matrix to extract appropriate elements. For example, to see those elements of x that satisfy both the conditions $(x > y)$ and $(x > 4)$, type x((x>y)&(x>4)).

In addition to these logical operators, there are many useful built-in logical functions, such as:

all	true $(= 1)$ if all elements of a vector are true,
	Example: all(x<0) returns 1 if each element of x is negative.
any	true $(= 1)$ if any element of a vector is true,
	Example: any(x) returns 1 if any element of x is nonzero.
exist	true $(= 1)$ if the argument (a variable or a function) exists,
isempty	true $(= 1)$ for an empty matrix,
isinf	true for all infinite elements of a matrix,
isfinite	true for all finite elements of a matrix,
isnan[3]	true for all elements of a matrix that are not a number (NaN), and
find	finds indices of nonzero elements of a matrix.
	Examples: find(x) returns [2 3 4] for x=[0 2 5 7] and
	[r,c]=find(A>100) returns the row and column indices
	i and j of A, in vectors r and c, for which $A_{ij} > 100$.

To complete this list of logical functions, we just mention isreal, issparse, isstr, and ischar.

3.2.4   Elementary math functions

For on-line help type:
help elfun

All of the following built-in math functions take matrix inputs and perform array operations (element by element) on them. Thus, they produce an output matrix of the same size as the input matrix.

Trigonometric functions

sin, sind	sine,	sinh	hyperbolic sine,
asin, asind	inverse sine,	asinh	inverse hyperbolic sine,
cos, cosd	cosine,	cosh	cyperbolic cosine,
acos, acosd	inverse cosine,	acosh	inverse hyperbolic cosine,
tan, tand	tangent,	tanh	hyperbolic tangent,

---

[3]The function isnan is the only way to check for NaNs in a matrix because any operation with a NaN produces a NaN and two NaNs are not equal to each other. Thus, find(A==nan) cannot find the indices of NaNs in matrix A. Of course, you could use some clever trick, such as find(A*0~=0).

atan, atand atan2	inverse tangent, four-quadrant $\tan^{-1}$,	atanh	inverse hyperbolic tangent,
sec, secd	secant,	sech	hyperbolic secant,
asec, asecd	inverse secant,	asech	inverse hyperbolic secant,
csc, cscd	cosecant,	csch	hyperbolic cosecant,
acsc, acscd	inverse cosecant,	acsch	inverse hyperbolic cosecant,
cot, cotd	cotangent,	coth	hyperbolic cotangent,
acot, acotd	inverse cotangent, and	acoth	inverse hyperbolic cotangent.

The angles given to these functions as arguments must be in *radians* for `sin`, `cos`, etc., and in *degrees* for `sind`, `cosd`, etc. Thus, `sin(pi/2)` and `sind(90)` produce the same result. All of these functions, except `atan2`, take a single scalar, vector, or matrix as input argument. The function `atan2` takes two input arguments, `atan2(y,x)`, and produces the four-quadrant inverse tangent such that $-\pi \le \tan^{-1}\frac{y}{x} \le \pi$. This gives the angle a rectangular to polar conversion.

Examples: If q=[0 pi/2 pi], x=[1 -1 -1 1], and y=[1 1 -1 -1], then

`sin(q)`	gives [0 1 0],
`sinh(q)`	gives [0 2.3013 11.5487],
`atan(y./x)`	gives [0.7854 -0.7854 0.7854 -0.7854], and
`atan2(y,x)`	gives [0.7854 2.3562 -2.3562 -0.7854].

Exponential functions

`exp`	exponential,
	Example: `exp(A)` produces a matrix with elements $e^{(A_{ij})}$.
	So how do you compute e^A? See the next section.
`log`	natural logarithm,
	Example: `log(A)` produces a matrix with elements $\ln(A_{ij})$.
`log10`	base 10 logarithm,
	Example: `log10(A)` produces a matrix with elements $\log_{10}(A_{ij})$.
`sqrt`	square root,
	Example: `sqrt(A)` produces a matrix with elements $\sqrt{A_{ij}}$.
	But what about $\sqrt{A}$? See the next section.
`nthroot`	real nth root of real numbers,
	Example: `nthroot(A,3)` produces a matrix with elements $\sqrt[3]{A_{ij}}$.

In addition, `log2`, `pow2`, `nextpow2`, `realpow`, `reallog`, `realsqrt`, `log1p` (for $\log(1+x)$), and `exp1m` (for $e^x - 1$) functions exist in MATLAB. Clearly, these are array operations. You can, however, also compute matrix exponential e^A, matrix square root $\sqrt{A}$, etc. See Section 3.2.5.

Complex functions

`abs`	absolute value, *Example:* `abs(A)` produces a matrix of absolute values $|A_{ij}|$.
`angle`	phase angle, *Example:* `angle(A)` gives the phase angles of complex A.
`complex`	constructs complex numbers from given real and imaginary parts,
	Example: `complex(A,B)` produces $A + Bi$.
`conj`	complex conjugate, *Example:* `conj(A)` produces a matrix with elements $\bar{A}_{ij}$.
`imag`	imaginary part, *Example:* `imag(A)` extracts the imaginary part of A.
`real`	real part, *Example:* `real(A)` extracts the real part of A.

Round-off functions

remember

`fix`	round toward 0,
	Example: `fix([-2.33 2.66])` $= [-2 \quad 2]$.
`floor`	round toward $-\infty$,
	Example: `floor([-2.33 2.66])` $= [-3 \quad 2]$.
`ceil`	round toward $+\infty$,
	Example: `ceil([-2.33 2.66])` $= [-2 \quad 3]$.
`mod`	modulus after division; `mod(a,b)` is the same as `a-floor(a./b)*b`,
	Example: `mod(26,5)` $= 1$ and `mod(-26,5)` $= 4$.
`round`	round toward the nearest integer,
	Example: `round([-2.33 2.66])` $= [-2 \quad 3]$.
`rem`	remainder after division, `rem(a,b)` is the same as `a-fix(a./b)*b`,
	Example: If `a=[-1.5 7]`, `b=[2 3]`, then `rem(a,b)` $= [-1.5 \ 1]$.
`sign`	signum function,
	Example: `sign([-2.33 2.66])` $= [-1 \quad 1]$.

3.2.5 Matrix functions

For on-line help type: `help matfun`

We discussed the difference between the array exponentiation `A.^2` and the matrix exponentiation `A^2` earlier. There are some built-in functions that are truly matrix functions and that also have array counterparts. The matrix functions are

`expm(A)`	finds the exponential of matrix A, e^A,
`logm(A)`	finds $\log(A)$ such that $A = e^{\log(A)}$, and
`sqrtm(A)`	finds $\sqrt{A}$.

The array counterparts of these functions are `exp`, `log`, and `sqrt`, which operate on each element of the input matrix (see Fig. 3.3 for examples). The matrix exponential function `expm` also has a specialized variant: `expm1`. See the on-line help or *MATLAB 7, Mathematics* [2] for their proper usage. MATLAB also provides a general matrix function `funm` for evaluating true matrix functions.

3.3 Character strings

For on-line help type: `help strfun`

All character strings are entered within two single right-quote characters—*'string'*. MATLAB treats every string as a row vector with one element per character. For example, typing

$$\text{message = 'Leave me alone'}$$

creates a vector, named `message`, of size 1×14 (spaces in strings count as characters). Therefore, to create a column vector of text objects, each text string must have exactly the same number of characters. For example, the command

$$\text{names = ['John'; 'Ravi'; 'Mary'; 'Xiao']}$$

creates a column vector with one name per row, although, to MATLAB, the variable `names` is a 4×4 matrix. Clearly, the command `howdy=['Hi'; 'Hello'; 'Namaste']` will result in an error because each row has a different length. Text strings of different lengths can be made to be of equal length by padding them with blanks. Thus, the correct input for `howdy` will be

```
>> A = [1 2; 3 4]

A =

     1      2
     3      4

>> asqrt = sqrt(A)

asqrt =

    1.0000     1.4142
    1.7321     2.0000
```

sqrt is an array operation. It gives the square root of each element of A as is evident from the output here.

```
>> Asqrt = sqrtm(A)

Asqrt =

   0.5537 + 0.4644i    0.8070 - 0.2124i
   1.2104 - 0.3186i    1.7641 + 0.1458i
```

sqrtm, on the other hand, is a true matrix function, i.e., it computes $\sqrt{A}$. Thus [Asqrt] * [Asqrt] = [A].

```
>> exp_aij = exp(A)

exp_aij =

    2.7183      7.3891
   20.0855     54.5982
```

Similarly, exp gives an element-by-element exponential of the matrix, whereas expm finds the true matrix exponential e^A. For information on other matrix functions, type help matfun.

```
>> exp_A = expm(A)

exp_A =

    51.9690     74.7366
   112.1048    164.0738
```

Figure 3.3: Examples of differences between matrix functions and array functions.

$$\text{howdy} = [\text{'Hi}_{\sqcup\sqcup\sqcup\sqcup\sqcup}\text{'}; \text{'Hello}_{\sqcup\sqcup}\text{'}; \text{'Namaste'}]$$

with each string being seven characters long ($\sqcup$ denotes a space).

An easier way of doing the same thing is to use the command `char`, which converts strings to a matrix. `char(s1,s2,...)` puts each string argument `s1`, `s2`, etc., in a row and creates a string matrix by padding each row with the appropriate number of blanks. Thus, to create the same `howdy` as previously, we type

$$\text{howdy} = \text{char('Hi','Hello','Namaste')}.$$

You can also use `strvcat` functions to create string arrays from string arguments. However, `char` is the best function for this purpose.

Because the same quote character (the single right quote) is used to signal the beginning as well as the end of a string, this character cannot be used inside a string for a quote or apostrophe. For a quote within a string you must use two single quotes, `''`. Thus, to title a graph with *3-D View of Boomerang's Path*, you write `title('3-D View of Boomerang''s Path')`.

3.3.1   Manipulating character strings

Character strings can be manipulated just like matrices. Thus,

$$\text{c} = [\text{howdy}(2,:)\quad \text{names}(3,:)]$$

produces `Hello Mary` as the output for c. This feature can be used along with number-to-string conversion functions, such as `num2str` and `int2str`, to create text objects containing dynamic values of variables. Such text objects are particularly useful in creating titles for figures and other graphics annotation commands (see Chapter 6). For example, suppose you want to produce a few variance-study graphs with different values of the sample size n, an integer variable. Producing the title of the graph with the command

```
title(['Variance study with sample size n = ',int2str(n)])
```

writes titles with the current value of n printed in the title.

There are several built-in functions for character string manipulation:

`char`	creates character arrays using automatic padding, also, converts ASCII numeric values to character arrays,
`abs`	converts characters to their ASCII numeric values,
`blanks`(n)	creates n blank characters,
`deblank`	removes the trailing blanks from a string,
`eval`	executes the string as a command (see Section 3.3.2),
`findstr`	finds the specified substring in a given string,
`int2str`	converts integers to strings (see example given earlier),
`ischar`	true ($= 1$) for character arrays,
`isletter`	true ($= 1$) for alphabetical characters,
`lower`	converts any uppercase letters in the string to lowercase,
`mat2str`	converts a matrix to a string,
`num2str`	converts numbers to strings (similar to `int2str`),
`strcmp`	compares two strings, returns 1 if same,
`strncmp`	compares the first n characters in given strings,
`strcat`	concatenates strings horizontally, ignoring trailing blanks,
`strvcat`	concatenates strings vertically, ignoring empty strings, and
`upper`	converts any lowercase letters in the string to uppercase.

The functions `char` and `strvcat` seem to do the same thing—create string arrays by putting each argument string in a row. So, what is the difference? `char` does not ignore empty strings but `strvcat` does. To see the difference, try the commands `char('up','','down')` and `strvcat('up','','down')`.

For on-line help type:
`help eval`

3.3.2   The `eval` function

MATLAB provides a powerful function `eval` to *evaluate* text strings and execute them if they contain valid MATLAB commands. For example, the command

$$\texttt{eval('x = 5*sin(pi/3)')}$$

computes $x = 5\sin(\pi/3)$ and is equivalent to typing `x=5*sin(pi/3)` on the command prompt.

There are many uses of the `eval` function. One use is in creating or loading sequentially numbered data files. For example, you can use `eval` to run a set of commands 10 times while you take a two-hour lunch break. The following script runs the set of commands 10 times and saves the output variables in 10 different files.

```
% Example of use of EVAL function
% A script file that lets you go out for lunch while MATLAB slogs
%----------------------
t = [0:0.1:1000];                    % t has 10001 elements
for k = 1:10
    outputfile = ['result',int2str(k)];  % see explanation below
    % write commands to run your function here
    theta = k*pi*t;
    x = sin(theta);                  % compute x
    y = cos(theta);                  % compute y
    z = x.*y.^2;                     % compute z
    % now save variables x, y, and z in a Mat-file
    eval(['save ',outputfile,' x y z'])  % see explanation below
end
```

The commands used here are a little subtle, so read them carefully. In particular, note that

- The first command, `outputfile=...`, creates a name by combining the counter of the `for` loop with the string `result`, thus producing the names `result1`, `result2`, `result3`, etc., as the loop counter k takes the values 1, 2, 3, etc.

- The last command `eval(...)` has *one* input argument—a long string that is made up of three strings: `'save '`, `outputfile`, and `' x y z'`. Note that whereas `save` and `x y z` are enclosed within quotes to make them character strings, the variable `outputfile` is not enclosed within quotes because it is already a string.

- The square brackets inside the `eval` command are used here to produce a single string by concatenating three individual strings. The brackets are not a part of the `eval` command. Note the blank spaces in the strings `'save␣'`

and '$\sqcup$x$\sqcup$y$\sqcup$z'. These spaces are essential. Can you see why? [Hint: Try to produce the string with and without the spaces.]

Similar to `eval` there is a function `feval` for evaluating functions. See discussion on `feval` in Section 4.2.2, page 107.

3.4 A Special Note on Array Operations

Although we have already discussed array operations in Section 3.2.1 on page 73, we highlight their importance by discussing them again here. Array operations are a big strength of MATLAB; they get rid of at least one `for` loop from serial computation. Most beginners have trouble with these operations.

Consider computing the product of data in two columns, x and y, where each column has n entries. That is, x and y are column vectors of length n. We need to compute xy for each entry. A natural way of computing this would be

```
for i=1:n
    xy(i) = x(i)*y(i);
end
```

This is a serial computation in a loop. In fact, if you had the data on a piece of paper in front of you, this is how you would calculate it by hand or by a regular calculator. This is how you would calculate the product in any formal computer language (possibly with some variations in the syntax of the commands). In MAT-LAB, however, these three lines of code can be reduced to one line by using the array operation on vectors x and y: `xy=x.*y`.

This is essentially what array operation is all about. The arrays considered here, vectors x and y, can be matrices too. In case of matrices, you get rid of two `for` loops—one for the row index and the other for the column index. Thus, the product of two 2-D arrays $A(n \times n)$ and $B(n \times n)$, involving the double-looped computation

```
for i=1:n
    for j=1:n
        C(i,j) = A(i,j)*B(i,j);
    end
end
```

can be replaced by, simply, `C=A.*B`. Please note that *this is not a matrix product* where $C_{ij} = \sum A_{ik} B_{kj}$. Here, we are computing $C_{ij} = A_{ij} B_{ij}$, just an element-by-element multiplication. As you already know, for matrix computations, you do not have to worry about the little dot; if A and B are compatible, then `C=A*B`. Much of the confusion regarding array operations arises from mixing matrix computations with array computations. Array computations are serial, element-by-element computations of two arrays.

Most built-in functions in MATLAB accept arrays as arguments. So, a `sinx` or `sqrtx` does not care whether x is a scalar, a 1-D array, or a 2-D array. However, if `x` is an array, then computing $\frac{\sin x}{x}$ or $x\sqrt{x}$ involves array operation on the two arrays. Thus, we need to compute `sin(x)./x` and `x.*sqrt(x)` or `x.^(3/2)`.

Addition or subtraction are, by definition, array operations and hence do not need special attention. A scalar enjoys special status and operates on scalars or arrays without any special treatment. Thus, for a scalar α, the computation αx is `alpha*x` whether x is a scalar or an array.

We end this discussion with a table of examples of typical computations with scalar and array arguments. In Table 3.1, variables x and y could be either a scalar or an array. Please look through each example and make sure that you understand when and why array operators (arithmetic operators preceded by a dot) are used.

Computation	Command with Scalar Arguments	Command with Vector Arguments
$x + 10$	`x+10`	`x+10`
$x/5$	`x/5`	`x/5`
$\sin x$	`sin(x)`	`sin(x)`
$\sqrt{x}$	`sqrt(x)`	`sqrt(x)`
$\frac{1}{x}$	`1/x`	`1./x`
x^2	`x^2`	`x.^2`
xy	`x*y`	`x.*y`
x^y	`x^y`	`x.^y`
$\sin x \cos x$	`sin(x)*cos(x)`	`sin(x).*cos(x)`
$\sin^2 x$	`sin(x)^2`	`sin(x).^2`
$\frac{e^x}{\sqrt{x}}$	`exp(x)/sqrt(x)`	`exp(x)./sqrt(x)`
e^{x^2}	`exp(x^2)`	`exp(x.^2)`
$\frac{1}{x^2+y^2}$	`1/(x^2+y^2)`	`1./(x.^2+y.^2)`

Table 3.1: Examples of computation involving array operations.

3.4.1 Vectorization

You should take full advantage of array operations by vectorizing your code or computation. By vectorization, we mean structuring your computation such that you can use vector or array variables with array operators instead of serial, scalar calculations. As an example, consider approximating the exponential function with the first 10 terms in its series expansion, $\exp(x) = \sum_{i=k}^{10} x^{k-1}/(k-1)!$. You can calculate it in a loop or a vectorized form:

```
x = 1;  k = [1:10];  e = sum(x.^(k-1)./factorial(k-1));
```

Vectorization is strongly recommended wherever possible.

3.5 Command-Line Functions

3.5.1 Inline functions

A mathematical function, such as $F(x)$ or $F(x, y)$, usually requires just the values of the independent variables for computing the value of the function. We frequently need to evaluate such functions in scientific calculations. One way to evaluate such functions is by programming them in function files (see Section 4.2 on page 102 for details). However, there is a quicker way of programming functions if they are not too complicated. This is done by defining *inline* functions—functions that are created on the command line. You define these functions using the built-in function `inline`.

For on-line help type:
`help inline`

The syntax for creating an inline function is particularly simple:

$$\text{F = inline('} function\ formula \text{')}$$

Thus, a function such as $F(x) = x^2 \sin(x)$ can be coded as

$$\text{F = inline('x^2 * sin(x)')}$$

However, note that our inline function can only take a scalar x as an input argument. We can modify it easily by changing the arithmetic operator to accept array argument: `F=inline('x.^2.*sin(x)')`. Once the function is created, you can use it as a function independently (e.g., type `F(5)` to evaluate the function at $x = 5$) or in the input argument of other functions that can evaluate it. See Fig. 3.4 for more examples and usage of inline functions. There are several examples of these functions in Section 3.8 on *plotting simple graphs*.

3.5.2 Anonymous functions

Anonymous functions are functions without names, created and referred by their *handles*. A function handle, created internally by MATLAB and stored in a user-defined variable, is basically the identity of the function.

An anonymous function is created by the command

$$\text{f = @(} input\ list \text{)}\ mathematical\ expression$$

where `f` is the function handle. The input list can contain a single variable or several variables separated by commas. After creating the function, you can use it with its handle to evaluate the function or pass it as an argument to other functions.

Examples:

`fx = @(x) x^2 - sin(x);`	creates a function $f(x) = x^2 - \sin x$,
`fx(5)`	evaluates $f(x)$ at $x = 5$,
`fxy = @(x,y) x^2 + y^2;`	creates a function $f(x, y) = x^2 + y^2$,
`fxy(2,3)`	evaluates $f(x, y)$ at $x = 2$ and $y = 3$,
`fx = @(x) x.^2 - sin(x)`	*vectorizes* the function $f(x)$, and
`x=[0:.1:pi/2]; plot(x,fx(x))`	plots $f(x)$ over 0 to $\pi/2$.

For a tutorial on anonymous functions, see Lesson 7 in Chapter 2. Anonymous functions are handy for command-line computation. For computing more involved functions that may require several intermediate calculations, you should use function files (see Lesson 5 in Chapter 2 and Section 4.2 on page 102).

```
>> Fx = inline('x^2 - sin(x)');          Create a simple inline function
>> Fx(pi)                                 F(x) = x² – sin(x) and compute
                                          its value at x = π.
ans =

        9.8696

>> Fx = inline('x.^2 - sin(x)');         Modify F(x) so that x can be an
>> x = [0 pi/4 pi/2 3*pi/4 pi];          array (i.e., use array operators).
>> Fx(x)                                  Evaluate F(x) when x is a vector.

ans =

        0      -0.0903    1.4674     4.8445     9.8696

>> sin(Fx(0.3))                           You can also use F(x) as an input
                                          argument of another function.
ans =

        -0.2041

>> x0 = fzero(Fx, pi/4)                    You can also use F(x) as an input
                                           argument to those functions that
x0 =                                       require a user-defined function.
                                           Here we use fzero to find a zero
        0.8767                             of F(x). See on-line help on fzero.

>> Fx(x0)

ans =

        2.2204e-16

>> Fxy = inline('(x.^2+y.^2)./exp(x.*y)');
>> check_sym = [Fxy(1,2)   Fxy(2,1)]
                                           Create another inline function
check_sym =                                F(x,y) of two variables. Note that
                                           here, F(x,y) is symmetric in x and y.
        0.6767         0.6767              Check its symmetry by evaluating
                                           F(x,y) at (1,2) and (2,1).
```

Figure 3.4: Examples of creating and using inline functions.

3.6 Using Built-in Functions and On-line Help

*For on-line help
type:*
`help help`

MATLAB provides hundreds of built-in functions for numerical linear algebra, data analysis, Fourier transforms, data interpolation and curve fitting, root-finding, numerical solution of ordinary differential equations, numerical quadrature, sparse matrix calculations, and general-purpose graphics. There is on-line help for all built-in functions. With so many built-in functions, it is important to know how to look for functions and how to learn to use them. There are several ways to get help:

`help` **the most direct on-line help:** If you know the exact name of a function, you can get help on it by typing `help` *functionname* on the command line. For example, typing `help help` provides help on the function `help` itself.

`lookfor` **the keyword search function:** If you are looking for a function, use `lookfor` *keyword* to get a list of functions with the string *keyword* in their description. For example, typing `lookfor 'identity matrix'` lists functions (there are two of them) that create identity matrices.

`helpwin` **the click and navigate help:** If you want to look around and get a feel for the on-line help by clicking and navigating through what catches your attention, use the window-based help, `helpwin`. To activate the help window, type `helpwin` at the command prompt or select Help Window from the Help menu on the command window menu bar.

`helpdesk` **the web browser–based help:** MATLAB provides extensive on-line documentation in both HTML and PDF formats. If you like to read on-line documentation and get detailed help by clicking on hyperlinked text, use the web browser–based help facility, `helpdesk`. To activate the help window, click on the help icon 🌐 on the menu bar. Alternatively, type `helpdesk` at the command prompt or select Product Help from the Help menu on the command window menu bar.

As you work more in MATLAB, you will realize that the on-line help with the command `help` and keyword search with `lookfor` are the easiest and fastest ways to get help.

Typing `help` by itself brings out a list of categories (see Fig. 3.5) in which help is organized. You can get help on one of these categories by typing `help` *category*. For example, typing `help elfun` gives a list of elementary math functions with a brief description of each function. Further help can be obtained on a function because the exact name of the function is now known.

On the other hand, `lookfor` is a friendlier command. It lets you specify a descriptive word about the function for which you need help. The following examples take you through the process of looking for help, getting help on the exact function that serves the purpose, and using the function in the correct way to get results.

```
>> help

HELP topics:
```

help by itself lists the names of categories in which the on-line help files are organized.

```
matlab\general      -  General purpose commands.
matlab\ops          -  Operators and special characters.
matlab\lang    ←    -  Programming language constructs.
matlab\elmat        -  Elementary matrices and matrix manipulation.
matlab\elfun        -  Elementary math functions.
matlab\specfun      -  Specialized math functions.
matlab\matfun       -  Matrix functions - numerical linear algebra.
matlab\datafun      -  Data analysis and Fourier transforms.
matlab\polyfun      -  Interpolation and polynomials.
matlab\funfun       -  Function functions and ODE solvers.
matlab\sparfun      -  Sparse matrices.
matlab\scribe       -  Annotation and Plot Editing.
matlab\graph2d      -  Two dimensional graphs.
matlab\graph3d      -  Three dimensional graphs.
matlab\specgraph    -  Specialized graphs.
matlab\graphics     -  Handle Graphics.
matlab\uitools      -  Graphical user interface tools.
matlab\strfun       -  Character strings.
matlab\imagesci     -  Image and scientific data input/output.
matlab\iofun        -  File input/output.
matlab\audiovideo   -  Audio and Video support.
matlab\timefun      -  Time and dates.
matlab\datatypes    -  Data types and structures.
matlab\verctrl      -  Version control.
matlab\codetools    -  Commands for creating and debugging code.
matlab\helptools    -  Help commands.
matlab\winfun       -  Windows Operating System Interface Files ...
matlab\demos        -  Examples and demonstrations.


>> help lang

Programming language constructs.
```

help category lists the functions in that category. Detailed help can then be obtained by typing: help functionname.

```
Control flow.

if        - Conditionally execute statements.
else      - Execute statement if previous IF condition failed.
elseif    - Execute if previous IF failed and condition is true.
end       - Terminate scope of control statements.
for       - Repeat statements a specific number of times.
 .          .
 .          .
 .          .
```

Figure 3.5: MATLAB help facility.

Caution: MATLAB's `help` command is not forgiving of any typos or misspellings, and hence you must know the exact command name.

3.6.1   Example 1: Finding the determinant of a matrix

We have a 10×10 matrix A (e.g., `A=rand(10)`) and we want to find its determinant. We do not know if the command for determinant is `det`, `deter`, or `determinant`. We can find the appropriate MATLAB command by searching with the `lookfor determinant` command, as shown in Fig. 3.6. Then we can find the exact syntax of the command or function. As is evident from the help on `det` in Fig. 3.6, all we have to do is type `det(A)` to find the determinant of A.

```
>> help determinant              To use help you must know the
determinant.m not found.         exact name of the function.


Use the help browser Search tab to search the documentation or
type "help help" for help command options such as help for methods.

>> lookfor determinant           To find the function, use the keyword
det          - Determinant.      search command lookfor.

>> help det
DET    Determinant.
    DET(X) is the determinant of the square matrix X.

    Use COND instead of DET to test for matrix singularity.


    See also cond.


    Overloaded methods
        help sym/det.m
    Reference page in Help browser
        doc det
```

Figure 3.6: Example of how to find the function that computes the determinant of a matrix.

3.6.2   Example 2: Finding eigenvalues and eigenvectors

Suppose we are interested in finding out the eigenvalues and eigenvectors of matrix A, but we do not know what functions MATLAB provides for this purpose. Because we do not know the exact name of the required function, our best bet to get help is to try `lookfor eigenvalue`. Figure 3.7 shows MATLAB's response to this command.

```
>> lookfor eigenvalue              lookfor provides keyword search for
                                    help files containing the search string.

A_eigenvalues     - A_eigenvalues            - rosser
wilkinson         - Wilkinson's eigenvalue test matrix.
balance           - Diagonal scaling to improve eigenvalue ...
condeig           - Condition number with respect to eigen...
eig               - Eigenvalues and eigenvectors.
ordeig            - Eigenvalues of quasitriangular matrices.
ordqz             - Reorder eigenvalues in QZ factorization.
ordschur          - Reorder eigenvalues in Schur factorization.
polyeig           - Polynomial eigenvalue problem.
qz                - QZ factorization for generalized eigen...
eigs              - Find a few eigenvalues and eigenvectors ...
eigshow           - Graphical demonstration of eigenvalues ...
expmdemo3         - Matrix exponential via eigenvalues and ...
mat4bvp           - Find the fourth eigenvalue of the ...
drum1             - One of the two model files, which ...
drum2             - One of the two model files, which ...
pdeeig            - Solve eigenvalue PDE problem.
pdeeigx           - Exact calculation of eigenvalues for ...
sptarn            - Solve generalized sparse eigenvalue problem
```

```
>> help eig                        Detailed help can then be obtained on
                                    any of the listed files with help.

  EIG   Eigenvalues and eigenvectors.
    E = EIG(X) is a vector containing the eigenvalues of
    a square matrix X.

    [V,D] = EIG(X) produces a diagonal matrix D of eigenvalues
    and a full matrix V whose columns are the corresponding
    eigenvectors so that X*V = V*D.

    [V,D] = EIG(X,'nobalance') performs the computation with
    balancing disabled, which sometimes gives more accurate
    results for certain problems with unusual scaling. If X is
    symmetric, EIG(X,'nobalance')is ignored since X is ...

    E = EIG(A,B) is a vector containing the generalized
    eigenvalues of square matrices A and B.

    [V,D] = EIG(A,B) produces a diagonal matrix D of
    generalized eigenvalues and a full matrix V whose columns
    are the corresponding eigenvectors so that A*V = B*V*D.
     .                 .
     .                 .
     .                 .
```

Figure 3.7: Example of use of on-line help.

As shown in Fig. 3.7, MATLAB lists all functions that have the string *eigenvalue* either in their name or in the first line of their description. You can then browse through the list, choose the function that seems closest to your needs, and ask for further help on it. In Fig. 3.7, for example, we seek help on function `eig`. The on-line help on `eig` tells us what this function does and how to use it.

Thus, if we are interested in just the eigenvalues of matrix A, we type `eig(A)` to get the eigenvalues, but if we want to find both the eigenvalues and the eigenvectors of A, we specify the output list explicitly and type `[eigvec,eigval]=eig(A)` (see Fig. 3.8). The names of the output or the input variables can be anything we choose. Although it's obvious, we note that the list of the variables (input or output) must be in the same order as specified by the function.

```
>> A = [5 -3  2; -3  8  4; 4  2  -9];
>> eig(A)


ans =

   -10.2206
     4.4246
     9.7960


>> [eigvec,eigval]  =  eig(A)


eigvec =

    0.1725      0.8706     -0.5375
    0.2382      0.3774      0.8429
   -0.9558      0.3156     -0.0247


eigval =

   -10.2206           0           0
         0      4.4246           0
         0           0      9.7960
```

Typing `eig(A)` without a list of outputs gives the eigenvalues of A in a column vector stored in the default output variable ans.

Specifying an explicit list of output variables `[eigvec,eigval]` gets the eigenvectors of A in the matrix `eigvec` and the eigenvalues of A on the diagonal of the matrix `eigval`.

Figure 3.8: Examples of use of the function `eig` to find eigenvalues and eigenvectors of a matrix.

Some comments on the help facility

- MATLAB is case-sensitive. All built-in functions in MATLAB use lowercase letters for their names, yet the **help** on any function lists the function in uppercase letters, as is evident from Figs. 3.5 and 3.7. For the first-time user, it may be frustrating to type the command exactly as shown by the help facility, i.e., in uppercase, only to receive an error message from MATLAB.

- On-line help can be obtained by typing help commands on the command line, as we have described in the previous section. This method is applicable on all platforms that support MATLAB. In addition, on some computers such as Macintosh and IBM compatibles with Windows, on-line help is also available on the menu bar under **Help**. The menu bar item gives options of opening **Help Window**, a point-and-click interactive help facility, or **Helpdesk**, a web browser–based extensive help facility.

- Although `lookfor` is a good command to find help if you are uncertain about the name of the function you want, it has an annoying drawback: it takes only one string as an argument. So typing `lookfor linear equations` causes an error (but `lookfor 'linear equations'` is ok), while both `lookfor linear` and `lookfor equations` do return useful help.

For on-line help type:
```
help general
help save
help load
help open
```

3.7   Saving and Loading Data

There are many ways of saving and loading data in MATLAB. The most direct way is to use the `save` and `load` commands. For a tutorial, see Lesson 9 in Chapter 2. You can also save a session or part of a session, including data and commands, using the `diary` command. We will describe `save` and `load` first.

3.7.1   Saving into and loading from the binary Mat-files

The `save` command can be used to save either the entire workspace or a few selected variables in a file called *Mat-file*. Mat-files are written in binary format with full 16-bit precision. It is also possible to write data in Mat-files in 8-digit or 16-digit ASCII format with optional arguments to the `save` command (see the on-line help). Mat-files must always have a .mat extension. The data saved in these files can be loaded into the MATLAB workspace by the `load` command. Examples of proper usage of these commands are as follows:

`save tubedata.mat x y`	saves variables x and y in the file **tubedata.mat**,
`save newdata rx ry rz`	saves variables *rx, ry*, and *rz* in the file **newdata.mat** (MATLAB automatically supplies the .mat extension to the file name),
`save xdata.dat x -ascii`	saves variable x in the file **xdata.dat** in 8-digit ASCII format,
`save`	saves the entire workspace in the file **matlab.mat**,
`load tubedata`	loads the variables saved in the file **tubedata.mat**,
`load`	loads the variables saved in the default file **matlab.mat**.

ASCII datafiles can also be loaded into the MATLAB workspace with the `load` command provided the datafile contains only a rectangular matrix of numbers. For more information, see the on-line help on `load`. To read and write ASCII files with specified delimiters (e.g., a tab), use `dlmread` and `dlmwrite`.

You can also use cut-and-paste between the MATLAB command window and other applications (such as Microsoft Excel) to import and export data.

3.7.2 Importing data files

For on-line help type:
`help uiimport`
`help importdata`
`help iofun`

MATLAB incorporates a good interface for importing different types of datafiles, both through GUI and the command-line. The easiest way to import data from a file is to invoke the *import wizard* by either (i) selecting **Import Data** from the **File** menu of the MATLAB window or (ii) typing `uiimport` on the command line and then following the instructions on the screen. The import wizard does an excellent job of recognizing most file formats, separating data into numeric data and text data, and loading the data in the MATLAB workspace. Import wizard can load data from several file formats used for text data, spreadsheet data (.xls, .wkl), movie data (.avi), image data (.tiff, .jpeg, etc.), and audio data (.wav, .au). See on-line help on `fileformats` for a list of readable file formats.

Alternatively, you can also import data from a file using the built-in function `importdata(filename)` from the command line. The file name must include the file extension (e.g., .xls for Excel files) so that `importdata` can understand the file format. See Lesson 9 in Chapter 2 for a tutorial.

3.7.3 Recording a session with `diary`

An entire MATLAB session, or part of one, can be recorded in a user-editable file by means of the `diary` command. A file name with any extension can be specified as the output file. For example, typing `diary session1.out` opens a diary file named `session1.out`. Everything in the command window, including user input, MATLAB output, error messages, etc., that follows the diary command is recorded in the file `session1.out`. Note that *a figure does not appear in the command window and is therefore not recorded in the diary*. The recording is terminated by the command `diary off`. The same diary file can be opened later during the same session by typing `diary on`. This will append the subsequent part of the session to the same file `session1.out`. All the figures in this book that show commands typed in the command window and consequent MATLAB output were generated with the `diary` command. Diary files may be opened and modified (say, to add some comments and clarifications) with any standard text editor.

This is the simplest method of recording commands and their output (sans figures), requires the least amount of effort, and produces a plain vanilla report. For producing nicely formatted reports, including commands, comments, numeric output, and figures, use the *publisher* (see Lesson 11 in Chapter 2, and Section 4.5 on page 129).

3.8   Plotting Simple Graphs

We close this section on interactive computation with an example of how to plot a simple graph and save it as an Encapsulated PostScript file.

As mentioned in the introduction, the plots in MATLAB appear in the graphics window. MATLAB provides very good facilities for both 2-D and 3-D graphics. The commands to produce simple plots are surprisingly simple. For complicated graphics and special effects there are a lot of built-in functions that enable the user to manipulate the graphics window in many ways. Unfortunately, the more control you want the more complicated it gets. We describe the graphics facility in more detail in Chapter 6.

The most direct command to produce a graph in 2-D is the `plot` command. If a variable `ydata` has n values corresponding to n values of variable `xdata`, then `plot(xdata,ydata)` produces a plot with `xdata` on the horizontal axis and `ydata` on the vertical axis. To produce overlay plots, you can specify any number of pairs of vectors as the argument of the `plot` command. We discuss more of this, and much more on other aspects of plotting, in Chapter 6. Figure 3.9 shows an example of plotting a simple graph of $f(t) = e^{t/10}\sin(t), \ \ 0 \leq t \leq 20$. This function could also be plotted using `fplot`, a command for plotting functions of a single variable. The most important thing to remember about the `plot` command is that the vector inputs for the x-axis data and the y-axis data must be of the same length. Thus, in the command `plot(x1,y1,x2,y2)`, $x1$ and $y1$ must have the same length, $x2$ and $y2$ must have the same length, but $x1$ and $x2$ may have different lengths.

The following are several other plotting functions that are easy to use and quickly produce plots.

fplot   It takes the function of a single variable and the limits of the axes as the input and produces a plot of the function. The simplest syntax (there is more to it, see the on-line help) is

> `fplot('function',[xmin xmax])`

Example: The following commands plot $f(x) = e^{-x/10}\sin(x)$ for x between 0 and 20 and produce the plot shown in Fig. 3.10.

```
fplot('exp(-.1*x).*sin(x)',[0, 20])
xlabel('x'),ylabel('f(x) = e^{x/10} sin(x)')
title('A function plotted with fplot')
```

In MATLAB, there is a suite of easy function plotters (their names begin with the prefix **ez**) that produce 2-D and 3-D plots and contour plots. These functions require the user to specify the function to be plotted as an input argument. The function to be plotted can be specified as a character string or as an inline function. There is a default domain for each **ez** plotter but you have the option of overwriting the default domain. These plotting functions are really easy to use. Take a quick look at the following examples.

```
>> x = 0: .1: 20;                           % create vector x
>> y = exp(0.1*x).*sin(x);                  % calculate y
>> plot(x,y)                                % plot x vs. y
>> xlabel('Time (t) in Seconds')           % label x-axis
>> ylabel('The Response Amplitude in mm')  % label y-axis
>> title('A Simple 2-D Plot')              % put a title
>> print resp_amp.eps -deps                % save the graph in
                                           % eps format in the
                                           % file resp_amp.eps
```

Filename   Device option

The output

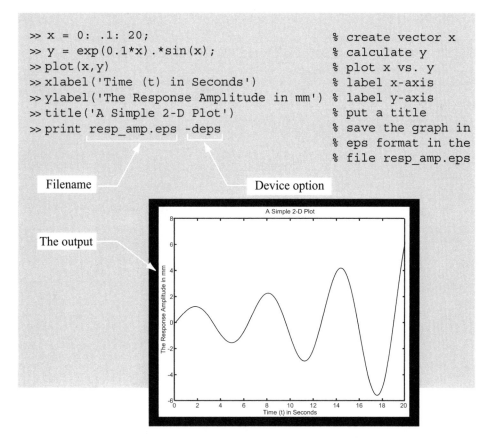

Figure 3.9: Example of a simple 2-D plot of function $f(t) = e^{t/10}\sin(t)$.

ezplot It takes the function of a single variable and the limits of the axes as the input and produces a plot of the function. The syntax is

> ezplot('*function*',[xmin, xmax])

where specification of the domain $x_{min} < x < x_{max}$ is optional. The default domain is $-2\pi < x < 2\pi$.

Example: The following command plots $f(x) = e^{-x/10}\cos(x)$ for x between 0 and 20 and produces the plot shown in Fig. 3.11.

> ezplot('exp(-.1*x).*cos(x)', [0, 20])

Alternatively, the commands F='exp(-.1*x).*cos(x)'; ezplot(F,[0,20]) and F=inline('exp(-.1*x).*cos(x)'); ezplot(F,[0,20]) will produce the same result. You can also use **ezplot** with implicit functions (i.e., you do not have an explicit formula for the function as $y = F(x)$, but you have an expression such as $x^2 y + \sin(xy) = 0$ or $F(x,y) = 0$). In addition, **ezplot** can also plot parametric curves given by $x(t)$ and $y(t)$, where $t_{min} < t < t_{max}$. See on-line help on **ezplot**.

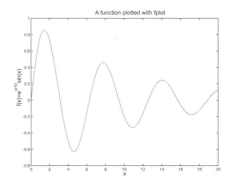

Figure 3.10: Example of plotting a function with `fplot`.

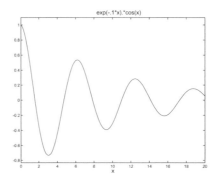

Figure 3.11: Example of plotting a function with `ezplot`.

ezpolar This is the polar version of `ezplot`. It takes $r(\theta)$ as the input argument. The default domain is $0 < \theta < 2\pi$.

Example: The following commands plot $r(\theta) = 1 + 2\sin^2(2\theta)$ for $0 < \theta < 2\pi$ and produce the plot shown in Fig. 3.12.

```
r = inline('1+2*(sin(2*t)).^2'); ezpolar(r)
```

ezplot3 takes $x(t)$, $y(t)$, and $z(t)$ as input arguments to plot 3-D parametric curves. The default domain is $0 < t < 2\pi$.

Example: The following commands plot $x(t) = t\cos(3\pi t)$, $y(t) = t\sin(3\pi t)$, and $z(t) = t$ over the default domain. The plot produced is shown in Fig. 3.13.

```
x = 't.*cos(3*pi*t)'; y = 't.*sin(3*pi*t)'; z = 't';
ezplot3(x,y,z)
```

Note that we can also specify x, y, and z as inline functions (see the next example).

ezcontour Contour plots are a breeze with this function. Just specify $Z = F(x,y)$ as the input argument and specify the domain for x and y if desired (other than the default $-2\pi < x < 2\pi$ and $-2\pi < y < 2\pi$).

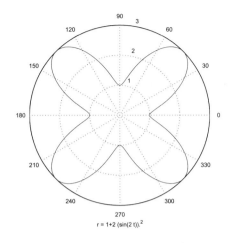

Figure 3.12: Polar plot of $r(\theta) = 1 + 2\sin^2(2\theta)$ with `ezpolar`.

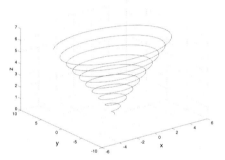

Figure 3.13: Plot of a 3-D parametric space curve with `ezplot3`.

Example: Let us plot the contours of $Z = \cos x \cos y \; \exp(-\sqrt{(x^2 + y^2)/4})$ over the default domain.

```
Z = inline('cos(x).*cos(y).*exp(-sqrt((x.^2+y.^2)/4))');
ezcontour(Z)
```

The plot produced is shown in Fig. 3.14(a).

ezcontourf This is just a variant of `ezcontour`. It produces filled contours. The syntax is the same as for `ezcontour`. As an example, let us take the same function as in `ezcontour` and change the default domain to $-5 < x < 5$ and $-5 < y < 5$: `ezcontourf(Z, [-5,5])`. The plot produced is shown in Fig. 3.14(b). For different ranges of x and y, the domain can be specified by a 1-by-4 vector, `[xmin, xmax, ymin, ymax]`.

ezsurf To produce stunning surface plots in 3-D, all you need to do is specify the function $Z = F(x, y)$ as an input argument of `ezsurf` and specify the domain of the function if you need to.

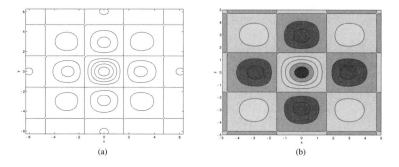

(a)   (b)

Figure 3.14: Contour plots of $Z = \cos x \cos y \, \exp(-\sqrt{(x^2 + y^2)/4})$ with (a) ezcontour and (b) ezcontourf.

Example: Here is an example of the surface plot for $Z = -5/(1 + x^2 + y^2)$ over the domain $|x| < 3$ and $|y| < 3$.

```
Z = inline('-5./(1+x.^2+y.^2)');
ezsurf(Z,[-3,3,-3,3])
```

The plot produced is similar to the one shown in Fig. 3.15 with no contours.

ezsurfc An even prettier looking variant of ezsurf is ezsurfc, which combines the surface plot with its contour plot. See Fig. 3.15, which is produced with the command ezsurfc(Z,[-3,3]) for the same function Z, as specified in the preceding ezsurf example.

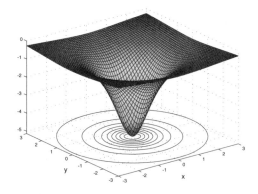

Figure 3.15: 3-D surface plot of $Z = -5/(1 + x^2 + y^2)$ with ezsurfc.

There are two more functions in the ez stable, ezmesh and ezmeshc, that are used for 3-D plotting. These functions work exactly like the surface plotting functions except that they create the 3-D graphs with wireframe meshes rather than surface patches.

EXERCISES

1. **See the structure of a matrix:** Create a 20×20 matrix with the command A=ones(20). Now replace the 10×10 submatrix between rows 6:15 and columns 6:15 with zeros. See the structure of the matrix (in terms of nonzero entries with the command spy(A). Set the 5×5 submatrices in the top right corner and bottom left corner to zeros and see the structure again.

2. **Create a symmetric matrix:** Create an upper triangular matrix with the following command:

$$A = \text{diag}(1:6) + \text{diag}(7:11,1) + \text{diag}(12:15,2).$$

Make sure you understand how this command works (see the on-line help on diag if required). Now use the upper off-diagonal terms of A to make A a symmetric matrix with the following command:

$$A = A + \text{triu}(A,1)'.$$

This is a somewhat *loaded* command. It takes the upper triangular part of A above the main diagonal, flips it (transposes), and adds to the original matrix A, thus creating a symmetric matrix A. See the on-line help on triu.

3. **Do some cool operations:** Create a 10×10 random matrix with the command A=rand(10). Now do the following operations.

   - Multiply all elements by 100 and then round off all elements of the matrix to integers with the command A=fix(A).
   - Replace all elements of A $<$ 10 with zeros.
   - Replace all elements of A $>$ 90 with infinity (inf).
   - Extract all $30 \leq a_{ij} \leq 50$ in a vector b, that is, find all elements of A that are between 30 and 50 and put them in a vector b.

4. **How about some fun with plotting?**

   - Plot the parametric curve $x(t) = t$, $y(t) = e^{-t/2} \sin t$ for $0 < t < \pi/2$ using ezplot.
   - Plot the cardioid $r(\theta) = 1 + \cos\theta$ for $0 < \theta < 2\pi$ using ezpolar.
   - Plot the contours of $x^2 + \sin(xy) + y^2 = 0$ using ezcontour over the domain $-\pi/2 < x < \pi/2$, $-\pi/2 < y < \pi/2$.
   - Create a surface plot along with contours of the function $H(x,y) = \frac{x^2}{2} + (1 - \cos y)$ for $-\pi < x < \pi$, $-2 < y < 2$.

5. **Automated saving and loading using multiple files:** (See Section 3.3.2)

   - Use a loop for generating magic matrices of size 20, 25, 30, 35, 40, 45, and 50.
   - In the same loop, generate a file name magicmatrix_i where i takes the value of 1, 2, ..., 7, as the loop index advances.
   - Use save and eval to save the generated magic matrix in the corresponding file, e.g., magic(20) should be saved in magicmatrix_1 and magic(35) should be saved in magicmatrix_4.
   - Clear the workspace of all variables. Use load and eval in another loop to load each of the seven files you just saved and to image the loaded matrix (use image(A) to image matrix A) in the figure window figure(i) where i is the loop index. [You could even save the figures in automatically generated files.]

4. Programming in MATLAB: Scripts and Functions

m file { Script / Function

A distinguishing feature of MATLAB is its *ease* of extendability through user-written programs. MATLAB provides its own language, which incorporates many features from C. In some regards, it is a higher-level language than most common programming languages, such as Pascal, Fortran, and C, meaning that you will spend less time worrying about formalisms and syntax. For the most part, MATLAB's language feels somewhat natural.

For on-line help type:
`help lang`

In MATLAB you write your programs in *M-files.* M-files are ordinary ASCII text files written in MATLAB's language. They are called M-files because they must have a .m at the end of their name (like myfunction.m). M-files can be created using any editor or word processing application.

There are two types of M-files—*script* files and *function* files. We will now discuss their purpose and syntax.

4.1 Script Files

A script file is an M-file with a set of valid MATLAB commands in it. A script file is executed by typing the name of the file (without the .m extension) on the command line. It is equivalent to typing all the commands stored in the script file, one by one, at the MATLAB prompt. Naturally, script files work on global variables, that is, variables currently present in the workspace. Results obtained from executing script files are left in the workspace.

A script file may contain any number of commands, including those that call built-in functions or functions written by you. Script files are useful when you have to repeat a set of commands several times. Here is an example.

Example of a script file: Let us write a script file to solve the following system of linear equations[1]:

$$\begin{bmatrix} 5 & 2r & r \\ 3 & 6 & 2r-1 \\ 2 & r-1 & 3r \end{bmatrix} \left\{ \begin{array}{c} x_1 \\ x_2 \\ x_3 \end{array} \right\} = \left\{ \begin{array}{c} 2 \\ 3 \\ 5 \end{array} \right\} \tag{4.1}$$

or $\mathbf{Ax} = \mathbf{b}$. Clearly, $\mathbf{A}$ depends on the parameter r. We want to find the solution of the equation for various values of the parameter r. We also want to find, say, the determinant of matrix $\mathbf{A}$ in each case. Let us write a set of MATLAB commands that do the job and store these commands in a file called solvex.m. How you create this file, write the commands in it, and save the file depends on the computer you are using. In any case, you are creating a file called solvex.m, which will be saved on some disk drive in some directory (or folder).

```
%----------- This is the script file 'solvex.m' ------------
% It solves equation (4.1) for x and also calculates det(A).

A = [5 2*r r; 3 6 2*r-1; 2 r-1 3*r];   % create matrix A
b = [2;3;5];                           % create vector b
det_A = det(A)                         % find the determinant
x = A\b                                % find x
```

In this example, we have not put a semicolon at the end of the last two commands. Therefore, the results of these commands will be displayed on the screen when we execute the script file. The results will be stored in variables det_A and x, and these will be left in the workspace.

Let us now execute the script in MATLAB.

```
>> clear all          % clear the workspace
>> r = 1;             % specify a value of r
>> solvex             % execute the script file solvex.m

det_A =

        64
x =
      -0.0312
       0.2344
       1.6875
>> who
```

This is the output. The values of the variables det_A and x appear on the screen because there is no semicolon at the end of the corresponding lines in the script file.

Check the variables in the workspace.

---

[1]If you are not familiar with matrix equations, see Section 5.1.1 on page 135.

Safe file under the script
The same title of the script
then give 'r' a value on the
command center and then type
title of script and Done!

You may notice that the value of r is assigned outside the script file and yet `solvex` picks up this value and computes A. This is because all the variables in the MATLAB workspace are available to script files, and all the variables in a script file are left in the MATLAB workspace. Even though the output of *solvex* is only *det_A* and x, A and b are also in your workspace, as you will see as a result of the last command `who`. So, if you want to do a big set of computations but in the end you want only a couple of variables as the output, a script file is not the right choice. What you need in this case is a *function file*.

Caution:

- *NEVER name a script file the same as the name of a variable it computes.* When MATLAB looks for a name, it first searches the list of variables in the workspace. If a variable of the same name as the script file exists, MATLAB will never be able to access the script file. Thus, if you execute a script file xdot.m that computes a variable *xdot*, then after the first execution the variable xdot exists in the MATLAB workspace. Now if you change something in the script file and execute it again, all you get is the old value of *xdot*! Your changed script file is not executed because it cannot be accessed as long as the variable xdot exists in the workspace. Fortunately, MATLAB detects this problem and gives an error message if the script file contains a variable with the same name as that of the script file.

- The name of a script file must begin with a letter. The rest of the characters may include digits and the underscore character. You may give long names but MATLAB will take only the first 19 characters. You may not use any periods in the name other than the last one in .m. Thus, names such as project_23C.m, cee213_hw5_1.m, and MyHeartThrobsProfile.m are fine, but project.23C.m and cee213_hw5.1.m are not valid names.

- Be careful with variable names while working with script files, because all variables generated by a script file are left in the workspace, unless you clear them. Avoid name clashes with built-in functions. It is a good idea to first check if a function or script file of the proposed name already exists. You can do this with the command `exist('proposed_name')`, which returns zero if nothing with the name *proposed_name* exists.

- You can create and test parts of a script without running the entire script by dividing the script into cells (computational units consisting of a few commands) and then running one cell script at a time from the *cell execution options* in the toolbar of the MATLAB editor window (see Section 4.5 on page 129 for a discussion of cell scripts). This is a MATLAB feature (introduced in MATLAB 7). The option to execute and test one cell at a time provides an excellent tool for developing long and complex programs without learning special debugging tools. The script cells we are talking about here should not be confused with *cells* that are one of the data types in MATLAB, discussed in Section 4.4.3.

4.2 Function Files

A function file is also an M-file, like a script file, except that the variables in a function file are all local. Function files are like programs or subroutines in Fortran, procedures in Pascal, and functions in C. Once you get to know MATLAB well, this is where you are likely to spend most of your time—writing and refining your own function files.

A function file begins with a function definition line, which has a well-defined list of inputs and outputs. Without this line, the file becomes a script file. The syntax of the function definition line is as follows:

> function [*output variables*] = *function_name(input variables)*;
>
> file name.

where the *function_name* must be the same as the file name (without the .m extension) in which the function is written. For example, if the name of the function is **projectile** it must be written and saved in a file with the name **projectile.m**. The function definition line may look slightly different, depending on whether there is no output, a single output, or multiple output.

Examples:

Function Definition Line   *File Name*
function [rho,H,F] = motion(x,y,t);   motion.m
function [theta] = angleTH(x,y);   angleTH.m
function theta = THETA(x,y,z);   THETA.m
function [] = circle(r);   circle.m
function circle(r);   circle.m

Caution: The first word in the function definition line, function, *must be typed in lowercase.* A common mistake is to type it as Function.

Anatomy of a function file

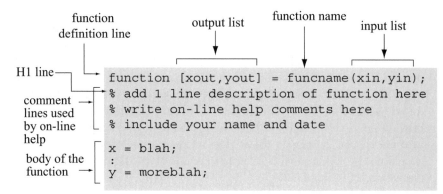

Features

- Comment lines start with a % sign and may be put anywhere. Anything after a % in a line is ignored by MATLAB as a nonexecutable statement.

- All comment lines immediately following the function definition line are displayed by MATLAB if **help** is sought on the function. The very first comment line immediately following the definition line is called the "H1" line. An H1 line, if present, is automatically cataloged in the **contents.m** file of the directory in which the file resides. This allows the line to be referenced by the **lookfor** command. A word of caution: any blanks before the % sign in the first comment line disqualify it from becoming an H1 line. Welcome to the idiosyncrasies of MATLAB!

- A single-output variable is not required to be enclosed in square brackets in the function definition line, but multiple output variables must be enclosed within [ ]. When there is no output variable present, the brackets as well as the equal sign may be omitted (see previous examples).

- Input variable names given in the function definition line are local to the function, so other variable names or values can be used in the function call. The name of another function can also be passed as an input variable. No special treatment is required for the function names as input variables in the function definition line. However, when the function is executed, the name of the input function must be passed as a character string, i.e., enclosed within two single right quotes (see example in the next section).

4.2.1 Executing a function

There are two ways a function can be executed, whether it is built-in or user-written:

1. **With explicit output:** This is the full syntax of calling a function. Both the output and input list are specified in the call. For example, if the function definition line of a function reads

   `function [rho,H,F] = motion(x,y,t);`

   then all the following commands represent legal call (execution) statements:

   - `[r,angmom,force]=motion(xt,yt,time);` The input variables *xt, yt,* and *time* must be defined before executing this command.
   - `[r,h,f]=motion(rx,ry,[0:100]);` The input variables *rx* and *ry* must be defined beforehand; the third input variable *t* is specified in the call statement.
   - `[r,h,f]=motion(2,3.5,0.001);` All input values are specified in the call statement.
   - `[radius,h]=motion(rx,ry);` Call with partial list of input and output. The third input variable must be assigned a default value inside the function if it is required in calculations. The output corresponds to the first two elements of the output list of **motion**.

2. **Without any output:** The output list can be omitted entirely if the computed quantities are not of any interest. This might be the case when the function displays the desired result graphically. To execute the function this way, just type the name of the function with the input list. For example, `motion(xt,yt,time);` will execute the function `motion` without generating any explicit output, provided that *xt, yt,* and *time* are defined. If the semicolon at the end of the call statement is omitted, the first output variable in the output list of the function is displayed in the default variable `ans`.

A function can be written to accept a partial list of inputs if some default values of the other unspecified inputs are defined inside the function. This kind of input list manipulation can be done with the built-in function `nargin`, which stands for number-of-arguments-in. Similarly, the list of outputs can be manipulated with the built-in function `nargout`. See the on-line help on `nargin` and `nargout`. For an example of how to use them, look at the function *fplot* by typing `type fplot.m`.

Example of a simple function file

Let us write a function file to solve the same system of linear equations that we solved in Section 4.1 using a script file. This time, we will make r an input to the function and *det_A* and x will be the output. Let us call this function `solvexf`. *As a rule, it must be saved in a file called* solvexf.m.

```
function [det_A, x] = solvexf(r);
% SOLVEXF solves a 3X3 matrix equation with parameter r
% This is the function file 'solvexf.m'
% To call this function, type:
%   [det_A,x] = solvexf(r);
% r is the input and det_A and x are output.
%_____

A = [5 2*r r; 3 6 2*r-1; 2 r-1 3*r]; % create matrix A
b = [2;3;5];                         % create vector b
det_A = det(A);                      % find the determinant
x = A\b;                             % find x.
```

Now r, x, and *det_A* are all local variables. Therefore, any other variable names may be used in their places in the function call statement. Let us execute this function in MATLAB.

```
>> clear all
>> [detA, y] = solvexf(1);   % take r=1 and execute solvexf.m
>> detA                      % display the value of detA
detA =
     64
>> y                         % display the value of y
y =
   -0.0312
    0.2344
    1.6875
>> who
Your variables are:
detA  y
```

Values of *detA* and *y* will be automatically displayed if the semicolon at the end of the function command is omitted.

Note that only *detA* and *y* are in the workspace; no *A*, *b*, or *x*.

After execution of a function, the only variables left in the workspace by the function will be the variables in the output list. This gives us more control over input and output than we can get with script files. We can also include error checks and messages inside the function. For example, we could modify the preceding function to check if matrix *A* is empty or not and display an appropriate message before solving the system by changing the last line to

```
if isempty(A)               % if matrix A is empty
    disp('Matrix A is empty');
else                        % if A is not empty
    x = A\b;                % find x
end                         % end of if statement.
```

For a description of `if-elseif-else` branching and other control-flow commands, see Section 4.3.4 on page 114.

4.2.2 More on functions

By now the difference between scripts and functions should be clear to you. The variables inside a function are local and are erased after execution of the function. But the variables inside a script file are left in the MATLAB workspace after execution of the script. Functions can have arguments, script files do not. What about functions inside another function? Are they local? How are they executed? Can a function be passed as an input variable to another function? Now we address these questions.

Executing a function inside another function

Usually, it is a straightforward process, so much so that you do not have to pay any special attention to it. In the function `solvexf`, we used a built-in function, `det`, to calculate the determinant of A. We used the function just the way we would use it at the MATLAB prompt or in a script file. This is true for all functions, built-in or user-written. The story is different only when you want the function name to be *dynamic*, that is, if the function to be executed inside may be different with different executions of the calling function.[2] In such cases, the actual name of the function is passed to the calling function through the input list and a dummy name is used inside the calling function. The mechanism of passing the function name and evaluating it inside the calling function is quite different from that for a variable. We explain it below.

A function in the input list

When a function needs to be passed in the input list of another function, the name of the function to be passed must appear as a character string in the input list. For example, the built-in function `fzero` finds a zero of a user-supplied function of a single variable (see Section 5.6, page 168, for more details). The call syntax of the function is `fzero(f,x)` where f is the name of the function or a *function handle* and x is an initial guess. There are several ways in which we can code the function f and pass it in the input list of `fzero`. We describe a few common ways of doing so on an example function:

$$f(r) = r^3 - 32r^2 + (r - 22)r + 100.$$

Use inline function: We can make $f(r)$ be an inline function (see Section 3.5.1 on page 83 for a detailed discussion) and pass its name to `fzero` as follows:

```
fr = inline('r^3 - 32*r^2 + (r-22)*r + 100');
r0 = fzero(fr,5);
```

Use function file: We can code $f(r)$ in a function file called funr.m as follows:

```
function f = funr(r);
% FUNR evaluates function f = r^3 -32 r^2+(r-22)r+100.
f = r^3 - 32*r^2 + (r-22)*r + 100;
```

Now we may call `fzero` with the statement:

```
r0 = fzero('funr',5).
```

Note the single quotes around the name of the input function `funr`.

Use function handle: A *function handle* is a convenient function identifier (a variable) created by the user. It is created with the @ symbol. We can use it for any function—built-in or a user-written M-file function (but not an inline function). As examples, let us do the same thing with `funr` as we did previously but with a function handle.

For on-line help type:
`help`
`function_handle`

---

[2]A function that uses another function inside its body is called a *calling function*. For example, `solvexf` in the example earlier is a calling function for the function `det`.

```
f1 = @funr          % create function handle f1 for 'funr'
r1 = fzero(f1,5);   % use function handle
```

Or, alternatively, `r1 = fzero(@funr,5)`.

Of course, anonymous functions, by definition, are created with explicit function handles and hence can be used directly in the input list.

Evaluating a function with `feval` or function handle

Functions can be evaluated indirectly (by reference) either by using their handle or using the function `feval`. Evaluating a function with its handle is straightforward as discussed earlier. It is just like using the actual functions except that we use the handle in place of the name of the function. For example, to evaluate the function `funr` created previously, we can use its function handle as follows:

```
hfun = @funr        % hfun is made the handle of funr
f = hfun(5)         % evaluate funr at r = 5
```

The function `feval` evaluates a function whose name is specified as a string in the list of input variables. For example,

For on-line help type:
`help feval`

$$[\text{y, z}] = \text{feval('Hfunction', x, t)};$$

evaluates the function `Hfunction` on the input variables x and t and returns the output in y and z. It is equivalent to typing `[y,z]=Hfunction (x,t)`. So why would you ever evaluate a function using `feval` when you can evaluate the function directly? The most common use is when you want to evaluate functions with different names but use the same input list. Consider the following situation. You want to evaluate any given function of x and y at the origin $x = 0$, $y = 0$. You can write a script file with one command in it:

$$\text{value} = \text{feval('funxy', 0, 0)};$$

This command is the same as writing `value=funxy(0,0)`. Now suppose Harry has a function $z(x,y) = \sin xy + xy^2$, programmed as

```
function  z = harrysf(x,y)
% function to evaluate z(x,y)
z = sin(x*y) + x*y^2;
```

and Kelly has a function $h(x,y) = 20xy - 3y^3 - 2x^3 + 10$, programmed as

```
function  h = kellysf(x,y)
% function to evaluate h(x,y)
h = 20*x*y - 3*y^3 - 2*x^3 + 10;
```

Both functions can be evaluated with your script file by changing the name `funxy` to `harrysf` and `kellysf`, respectively. The point here is that the command in your script file takes *dynamic* filenames.

Of course, the same can be done using a function handle too. Let us say the function handle is `hf`. Then the function evaluation at $x = 0$ and $y = 0$ is done with `value = hf(0,0)`. To evaluate `harrysf`, we assign `hf = @harrysf`, and to evaluate `kellysf`, we assign `hf = @kellysf` before using `value = hf(0,0)`.

The indirect evaluation of a function (by reference) becomes essential when a function is passed as an input variable to another function. In such cases, the function passed as the input variable must be evaluated using either its function handle or `feval`. For example, the ODE solvers `ode23` and `ode45` take user-defined functions as inputs, in which the user specifies the differential equation. Inside `ode23` and `ode45`, the user-defined function is evaluated at the current time t and the current value of x to compute the derivative $\dot{x}$ using `feval`. `ode23` and `ode45` are M-files, which you can copy and edit. Make a printout of one of them and see how it uses `feval` and function handle.

Writing good functions

Writing functions in MATLAB is easier than writing functions in most standard programming languages, or, for that matter, in most of the software packages that support their own programming environment. However, writing efficient and elegant functions is an art that comes only through experience. For beginners, keeping the following points in mind helps.

- **Pseudocode:** Before you begin writing a function, write a *pseudocode*. It is essentially the entire function coded in English (as opposed to the programming language) with clear control flow. Think about the logical structure and sequence of computations, define the input and output variables, and write the function in plain words. Then begin the translation into MATLAB language.

- **Readability:** Select a sensible name for the function and the variables inside it. Write enough comments in the body of the function. Design and write helpful comments for on-line help (the comment lines in the beginning of the function). Make sure you include the syntax of how to use the function.

- **Modularity:** Keep your functions *modular*, that is, break big computations into smaller chunks and write separate functions for them. Keep your functions small in length.

- **Robustness:** Provide checks for errors and exit with helpful error messages.

- **Expandability:** Leave room for growth. For example, if you are writing a function with scalar variables but think you may use vector variables later, write the function with this in mind. Avoid hard-coding actual numbers.

4.2.3   M-Lint code analyzer

When you write a program in MATLAB, and create a script or a function, you want to make sure that your program

- uses correct syntax for each statement,
- has proper function definition line if it is a function,
- uses appropriate built-in functions, and
- contains no unresolvable references.

Even though it is relatively easy to take care of these things if a function is just a few lines long, it becomes harder as the length of the code increases. It used to be that the only way you found your mistakes was when you executed the function and got error messages. Now, MATLAB provides an assistant to help you in this task. It is called the *M-Lint Automatic Code Analyzer* (offers automatic corrections too). It is an excellent facility for helping you in developing error free codes. There are two basic ways in which you can use M-Lint code analyzer:

1. **On fresh code as you write it:** When you open a new M-file in the MATLAB editor using File→New→Blank M-File or Function M-File from the MATLAB menu, M-Lint code analyzer is pressed into service automatically. As you write the lines of code, M-Lint, as a nice assistant, starts working quietly, watching over your shoulders, and lists its objections politely and symbolically in the right-hand margin of the editor window. There is a colored small square on the top (in the right margin) that indicates the level of M-Lint's happiness with your code—a red-faced square indicates error, a green-faced square is a signal to march on. Below the square, there may be orange- or red-colored lines corresponding to a particular line of code. Place your cursor on these colored lines (or, alternatively, on the underlined items in your code) one by one to see the message your assistant has left for you while it checked the line. Orange lines contain advisories (warnings) but red lines must necessarily be attended to. Many a times, fixing one error gets rid of many other warning lines too.

2. **On existing M-files:** You can open an existing M-file in the editor and see M-Lint's messages just the way you would on a new M-file. Alternatively, you can run M-Lint on the whole directory and produce reports for each M-file in the directory with a single click—go to the Current Directory pane, click on the *Action* icon (the little gear icon in the menu bar of the pane), select Reports→M-Lint Code Check Report from the pop-up menu. You are presented with the M-Lint report for all M-files. There is a lot more you can do, customize reports, set your M-Lint preferences, etc. See the on-line documentation to learn more if you wish.

Use M-Lint as a friendly assistant. Listen to its advice and make sure that it is happy (green-faced square) before you close your M-file.

4.2.4 Subfunctions

MATLAB allows several functions to be written in a single file. Although this facility is useful from a file organization point of view, it comes with severe limitations. All functions written below the first function in the file are treated as subfunctions and are NOT available to the outside world. This means that whereas the first function can access all its subfunctions, and the subfunctions written in the same file can also access each other, functions outside this file cannot access these subfunctions. You can, however, get on-line help on a subfunction by typing

```
help m_filename>subfunction_name
```
.

4.2.5   Nested functions:

Nested functions are functions written inside a main function, just like subfunctions but with the following important distinctions:

- Each nested function must be terminated by an **end** statement. For example:

```
function [x, y] = main_fun(t, a, b)
   :
     function x = nested_fun1(a,b)
     :
     end
     function y = nested_fun2(t)
     :
     end
end
```

  Here, nested_fun1 and nested_fun2 are nested functions inside the main function main_fun. Note that the main function (whose name must be the same as the M-file name) must be terminated by an **end** too.

- All nested functions share the workspace of functions in which they are nested. Thus nested_fun1 and main_fun share their workspace variables and so do nested_fun2 and main_fun, but nested_fun1 and nested_fun2 do not share workspace variables. This facility of sharing workspace makes it easy for the nested functions to access each other's variables and their values without any explicit declaration (e.g., global) or passing them in the input list.

Functions can be nested to any level; that is, nested functions can also have their own nested functions. Of course, nested functions are not visible or accessible from outside the main function. They can, however, be made accessible from outside by creating their explicit function handles (see on-line help on nested functions in the help browser).

4.2.6   Compiled (parsed) functions: The p-code

When a function is executed in MATLAB, each command in the function is first interpreted by MATLAB and then translated into the lower-level language. This process is called *parsing*. It is not exactly like compiling a function in C or Fortran but, superficially, it is a similar process. MATLAB allows you to save a parsed function for later use. To parse a function called projectile that exists in an M-file called projectile.m, type the command

```
                    pcode projectile
```

This command generates a parsed function that is saved in a file named projectile.p. Next time you call the function projectile, MATLAB directly executes the preparsed function.

For all moderate-size functions, saving them as parsed functions does not save as much time during execution as you would think. The MATLAB parser is quick enough to reduce the overhead on parsing to a negligible fraction. The best use of p-codes, perhaps, is in protecting your proprietary rights when you send your functions to other people. When you send p-codes, the recipients can execute them but they cannot modify them.

4.2.7 The profiler

For on-line help type:
`help profile`

To evaluate the performance of functions, MATLAB provides a tool called *profiler*. The profiler keeps track of the time spent on each line of the function (in units of 0.001 seconds) as the function is executed. The profile report shows how much time was spent on which function and which line. The report can be generated in various forms and levels of detail, depending on the options used.

The profiler is surprisingly easy to use. Let us see a pseudoexample. To profile the function solvexf, type the following commands:

```
profile on          % turn the profiler on
[d,x] = solvexf(3)  % execute the function solvexf
profile viewer      % invoke the profile viewer to see results
profile off         % thank you, bye profiler
```

The profile viewer brings up a pop-up window that shows the report in HTML format. Locate your function in the list and click on it to see a line-by-line report on your function.

Note: The profiler implemented in MATLAB is fairly advanced. You can use it to produce and display various statistics about your programs, including the functions (written by you or built-in) that your program calls. This is not something you should spend your time on if you are a beginner.

4.3 Language-specific Features

For on-line help type:
`help lang`

We have already discussed numerous features of MATLAB's language through many examples in the previous sections. You are advised to pay special attention to proper usage of punctuation marks and different delimiters (Appendix A) and operators, especially the array operators (a period (.) preceding the arithmetic operators, Section 3.2.1) and the relational operators (Section 3.2.2). For control-flow, MATLAB provides `for` and `while` loops, an `if-elseif-else` construct, and a `switch-case-otherwise` construct. All the control-flow statements must terminate with corresponding `end` statements. We now discuss flow control and some other specific features of the language. See the on-line help for more details.

4.3.1 Use of comments to create on-line help

As we have already pointed out in the discussion on function files (Section 4.2, page 102), the comment lines at the beginning (before any executable statement) of a script or a function file are used by MATLAB as the on-line help on that file. This automatically creates the on-line help for user-written functions. It is a good idea to copy the function definition line without the word `function` among those first few comment lines so that the execution syntax of the function is displayed by the on-line help. The command `lookfor` looks for the argument string in the first commented line of M-files. Therefore, in keeping with the somewhat confusing convention of MATLAB's built-in functions, you should write the name of the script or function file in uppercase letters, followed by a short description with keywords as the first commented line. Thus, the first line following the function definition line in the example function on page 104, Section 4.2.1, reads

```
% SOLVEXF solves a 3x3 matrix equation with parameter r.
```

4.3.2 Continuation

An ellipsis (three consecutive periods) at the end of a line denotes continuation. So, if you type a command that does not fit on a single line, you may split it across two or more lines by using an ellipsis at the end of each but the last line.

Examples:

```
A = [1 3 3 3; 5 10 -2 -20; 3 5 ...
      10 2; 1 0 0 9];
x = sin(linspace(1,6*pi,100)) .* cos(linspace(1,6*pi,100)) +...
      0.5*ones(1,100);
plot(tube_length,fluid_pressure,':',tube_length,...
      theoretical_pressure,'-')
```

(The last command assumes that variables `tube_length`, `fluid_pressure`, etc. exist in the workspace.)

You *may not*, however, use continuation inside a character string. For example, typing

```
logo = 'I am not only the President and CEO of Miracle Hair,...
        but also a client';
```

produces an error. For creating such long strings, break the string into smaller string segments and use concatenation (see Section 3.3).

4.3.3 Global variables

It is possible to declare a set of variables to be globally accessible to all or some functions without passing the variables in the input list. This is done with the `global` command. For example, the statement `global x y z` declares the variables x, y, and z to be global. This statement goes before any executable statement in the functions and scripts that need to access the values of the global variables. Be careful with the names of the global variables. It is generally a good idea to name such variables with long strings to avoid any unintended match with other local variables.

Example: Consider solving the following first-order ODE:

$$\dot{x} = kx + c\sin t, \qquad x(0) = 1.0$$

where you are interested in solutions for various values of k and c. Your script file may look like

```
% scriptfile to solve a first-order ode.
% -------------
ts = [0  20];                  % specify time span=[t_0 t_final]
x0 = 1.0;                      % specify initial condition
[t, x] = ode23 ('ode1',ts,x0); % execute ode23 to solve the ODE.
plot(t,x)                      % plot the result
```

and the function ode1 may look like

```
function xdot = ode1(t,x);
% ODE1: function to compute the derivative xdot
% at given t and x.
% Call syntax: xdot = ode1 (t,x);
% -------------
xdot = k*x + c*sin(t);
```

This, however, won't work. In order for ode1 to compute xdot, the values of k and c must be prescribed. These values could be prescribed inside the function ode1 but you would have to edit this function each time you change the values of k and c. An alternative[3] is to prescribe the values in the script file and make them available to the function ode1 through *global* declaration.

```
% scriptfile to solve a first-order ode.
% -------------
global k_value c_value         % declare global variables
k_value = 5;   c_value = 2;     % specify their values
ts = [0 20];                    % specify time span
x0 = 1.0;                       % specify initial conditions
[t, x] = ode23('ode1',ts,x0);  % execute ode23 to solve the ODE
plot(t,x)                       % plot the result
```

Now you have to modify the function ode1 so that it can access the global variables:

```
function xdot = ode1(t,x);
% ODE1: function to compute the derivative xdot
%          at given t and x.
% Call syntax: xdot = ode1(t,x);
% -------------
global k_value c_value
xdot = k_value*x + c_value*sin(t)
```

Now, if the values of *k_value* and *c_value* are changed in the script file, the new values become available to ode1 too. Note that the global declaration is only in the script file and the user function file ode1, and therefore *k_value* and *c_value* will be available to these files only.

---

[3]Another alternative is to use ode23 to pass the variables k and c to ode1. See Section 5.5.5 on page 163 for details.

4.3.4 Loops, branches, and control-flow

MATLAB has its own syntax for control-flow statements such as `for` loops, `while` loops and, of course, `if-elseif-else` branching. In addition, it provides three commands—`break`, `error`, and `return`—to control the execution of scripts and functions. A description of these functions follows.

For loops

A `for` loop is used to repeat a statement or a group of statements for a fixed number of times. Here are two examples:

Example 1:
```
for m=1:100
      num = 1/(m+1)
end
```
Example 2: `for n=100:-2:0, k = 1/(exp(n)), end`

The *counter* in the loop can also be given explicit increment: `for i=m:k:n` to advance the counter i by k each time (in the second example, n goes from 100 to 0 as 100, 98, 96, ..., etc.). You can have nested `for` loops, that is, `for` loops within `for` loops. *Every* `for`, *however, must be matched with an* `end`.

While loops

A `while` loop is used to execute a statement or a group of statements for an indefinite number of times until the condition specified by `while` is no longer satisfied. For example:
```
% let us find all powers of 2 below 10000
v = 1;   num = 1;  i=1;
while num < 10000
      v = [v; num];
      i = i + 1;
      num = 2^i;
end
v       % display v
```
Once again, a `while` must have a matching `end`.

If-elseif-else statements

This construction provides a logical branching for computations. Here is an example:
```
i=6; j=21;
if i > 5
   k = i;
elseif (i>1) & (j==20)
   k = 5*i + j;
else
```

```
        k = 1;
    end
```

Of course, you can have nested **if** statements as long as you have matching **end** statements. You can nest all three kinds of loops, in any combination.

Switch-case-otherwise

This construction provides another logical branching for computations. A flag (any variable) is used as a *switch* and the values of the flag make up the different *cases* for execution. The general syntax is:

```
switch flag
case value1
        block 1 computation
case value2
        block 2 computation
otherwise
        last block computation
end
```

Here, *block 1 computation* is performed if **flag=value1**, *block 2 computation* is performed if **flag=value2**, and so on. If the value of **flag** does not match any **case**, then the *last block computation* is performed. Of course, like all good things in life, the **switch** must come to an **end** too.

The **switch** can be a numeric variable or a string variable. Let us look at a more concrete example using a string variable as the **switch**:

```
color = input('color = ','s');
switch color
case 'red'
        c = [1 0 0];
case 'green'
        c = [0 1 0];
case 'blue'
        c = [0 0 1];
otherwise
        error('invalid choice of color')
end
```

Break

The command **break** inside a **for** or **while** loop terminates the execution of the loop, even if the condition for execution of the loop is true.

Examples: (Assume that the variables used in the codes below are predefined.)

```
1.    for i=1:length(v)
          if v(i) < 0          % check for negative v
              break            % terminate loop execution
          end
          a = a + v(i);        % do something
      end
2.    x = exp(sqrt(163));
      while 1
          n = input('Enter max. number of iterations ')
          if n <= 0
              break                % terminate loop execution
          end
          for i=1:n
              x = log(x);      % do something
          end
      end
```

If the loops are nested then break terminates only the innermost loop.

Error

The command error('*message*') inside a function or a script aborts the execution, displays the error message *message*, and returns the control to the keyboard.
Example:

```
function c = crossprod(a,b);
% crossprod(a,b) calculates the cross product axb.
if nargin~=2          % if not two input arguments
   error('Sorry, need two input vectors')
end
if length(a)==2       % begin calculations
   ....
end
```

Return

The command return simply returns the control to the invoking function.
Example:

```
function animatebar(t0,tf,x0);
% animatebar animates a bar pendulum.
:
disp('Do you want to see the phase portrait?')
ans = input('Enter 1 if YES, 0 if NO ');
                            % see text for description
if ans==0                   % if the input is 0
   return                   % exit function
```

```
else
   plot(x,...)              % show the phase plot
end
```

4.3.5   Interactive input

For on-line help
type:
help lang

The commands `input`, `keyboard`, `menu`, and `pause` can be used inside a script or function file for interactive user input. Their descriptions follow.

Input

The command `input('`*`string`*`')`, used in the previous example, displays the text in *string* on the screen and waits for the user to give keyboard input.

Examples:

- `n=input('Largest square matrix size ')`; prompts the user to input the size of the "largest square matrix" and saves the input in `n`.
- `more=input('More simulations?  (Y/N)','s')`; prompts the user to type Y for YES and N for NO and stores the input as a string in `more`. Note that the second argument, `'s'`, of the command directs MATLAB to save the user input as a string.

This command can be used to write *user-friendly* interactive programs in MATLAB.

Keyboard

The command `keyboard` inside a script or a function file returns control to the keyboard at the point where the command occurs. The execution of the function or the script is *not* terminated. The command window prompt ≫ changes to k≫ to show the special status. At this point, you can check variables already computed, change their values, and issue any valid MATLAB commands. The control is returned to the function by typing the word `return` on the special prompt k≫ and then pressing the return/enter key.

This command is useful for debugging functions. Sometimes, in long computations, you may like to check some intermediate results, plot them and see if the computation is headed in the right direction, and then let the execution continue.

Example:

```
% EXKEYBRD: a script file for example of keyboard command

A = ones(10)              % make a 10x10 matrix of 1s
for i=1:10
    disp(i)              % display the value of i
    A(:,i) = i*A(:,i);   % replace the ith column of a
    if i==5             % when i = 5
        keyboard        % return the control to keyboard
    end
end
```

During the execution of the preceding script file **exkeybrd.m**, the control is returned to the keyboard when the value of the counter i reaches five. The execution of **exkeybrd** resumes after the control is returned to the script file by typing **return** on the special prompt k≫.

Menu

The command menu(*'MenuName'*, *'option1'*, *'option2'*, ...) creates an on-screen menu with the *MenuName* and lists the options in the menu. The user can select any of the options using the mouse or the keyboard, depending on the computer. The implementation of this command on Macs and PCs creates a nice window menu with buttons.
Example:

```
% Plotting a circle
r = input('Enter the desired radius ');
theta = linspace(0,2*pi,100);
r = r*ones(size(theta));    % make r the same size as theta
coord = menu('Circle Plot','Cartesian','Polar');
if coord==1                 % if the first option is selected
                            %- from the menu
    plot(r.*cos(theta),r.*sin(theta))
    axis('square')
else                        % if the second option is selected
                            %- from the menu
    polar(theta,r);
end
```

In this script file, the **menu** command creates a menu with the name **Circle Plot** and two options—**Cartesian** and **Polar**. The options are internally numbered. When the user selects one of the options, the corresponding number is passed on to the variable *coord.* The **if-else** construct following the **menu** command shows what to do with each option. Try out this script file.

Pause

The command **pause** temporarily halts the current process. It can be used with or without an optional argument:

pause   halts the current process and waits for the user to give a
            "go-ahead" signal. Pressing any key resumes the process.
            Example: for i=1:n, plot(X(:,i),Y(:,i)), pause, end.

pause(n) halts the current process, pauses for n seconds,
            and then resumes the process.
            Example: for i=1:n, plot(X(:,i),Y(:,i)), pause(5), end
                   pauses for five seconds before it plots the next graph.

4.3.6   Recursion

The MATLAB programming language supports recursion, i.e., a function can call itself during its execution. Thus, recursive algorithms can be directly implemented in MATLAB (what a break for Fortran users!). To illustrate this, let us look at a simple example of computing the nth term in the Fibonacci sequence (actually, first discovered by Pingala (300–200 BCE), an ancient Indian mathematician): $0, 1, 1, 2, 3, 5, 8, \ldots$. If we label the terms as F_0, F_1, F_2, etc., then the recursion relationship for generating this sequence is: $F_k = F_{k-1} + F_{k-2}$ for $k > 2$. The seeds are $F_0 = 0$ and $F_1 = F_2 = 1$. The nth term in this sequence can be computed by the following recursive function:

```
function Fn = fibonacci(n)
% FIBBONACI: computes nth term in the Fibonacci sequence
% written by Abhay, May 15, 09, modified by RP, June 1, 09
if n==0, Fn = 0;                            % Fn=0 for n=0
   elseif n==1 | n==2, Fn = 1;              % Fn=1, for n=1 OR n=2
   else Fn = fibonacci(n-1) + fibonacci(n-2); % recursion relation
end
```

As you can see, this function calls itself when computing F_n.

4.3.7   Input/output

MATLAB supports many standard C-language file I/O functions for reading and writing formatted binary and text files. The functions supported include

For on-line help type:
`help iofun`

`fopen`	opens an existing file or creates a new file,
`fclose`	closes an open file,
`fread`	reads binary data from a file,
`fwrite`	writes binary data to a file,
`fscanf`	reads formatted data from a file,
`fprintf`	writes formatted data to a file,
`sscanf`	reads strings in specified format,
`sprintf`	writes data in formatted string,
`fgets`	reads a line from a file including new-line character,
`fgetl`	reads a line from a file discarding new-line character,
`frewind`	rewinds a file,
`fseek`	sets the file position indicator,
`ftell`	gets the current file position indicator, and
`ferror`	inquires file I/O error status.

You are likely to use only the first six commands in the list for file I/O. For most purposes, `fopen`, `fprintf`, and `fclose` should suffice. For a complete description of these commands, see the on-line help or consult *MATLAB 7 Programming* [3] or a C-language reference book [8].

Here is a simple example that uses fopen, fprintf, and fclose to create and write formatted data to a file:

```
% TEMTABLE - generates and writes a temperature table
% Script file to generate a Fahrenheit-Celsius
% temperature table. The table is written in
% a file named 'Temperature.table'.
% ---------------------------------------------------
F=-40:5:100;            % take F=[-40 -35 -30 .. 100]
C=(F-32)*5/9;           % compute corresponding C
t=[F;C];                % create a matrix t (2 rows)
fid = fopen('Temperature.table','w');
fprintf(fid,'  Temperature Table\n ');
fprintf(fid,' ~~~~~~~~~~~~~~~~ \n');
fprintf(fid,'Fahrenheit    Celsius \n');
fprintf(fid,'  %4i      %8.2f\n',t);
fclose(fid);
```

In this script file, the first I/O command, fopen, opens a file **Temperature.table** in the *write* mode (specified by 'w' in the command) and assigns the *file identifier* to fid. The following fprintf commands use fid to write the strings and data to that file. The data is formatted according to the specifications in the string argument of fprintf. In the previous command, \n stands for *new line*, %4i stands for an *integer* field of width 4, and %8.2f stands for a *fixed point* field of width 8 with two digits after the decimal point.

The output file, Temperature.table, is shown here. Note that the data matrix t has two rows, whereas the output file writes the matrix in two columns. This is because t is read columnwise and then written in the format specified (two values in a row).

```
     Temperature Table
    ~~~~~~~~~~~~~~~~

  Fahrenheit    Celsius
     -40        -40.00
     -35        -37.22
     -30        -34.44
     -25        -31.67
      :           :
      75         23.89
      80         26.67
      85         29.44
      90         32.22
      95         35.00
     100         37.78
```

4.4   Advanced Data Objects

MATLAB supports several data objects other than by-now-familiar 2-D arrays. Two among these are *structures* and *cells*. They have fundamentally different data structures. Also, the most familiar data object, a matrix, can be *multidimensional*. Although a detailed discussion of these objects and their applications is beyond the scope of this book, this section gives enough introduction and examples to get you started. At first, these objects look very different from a matrix, MATLAB's heart and soul. The fact, however, is that these objects, as well as a matrix, are just special cases of the fundamental data type, the *array*. As you become more comfortable with these objects, they fall into their right places and you start working with them just as you would with vectors and matrices.

4.4.1   Multidimensional matrices

MATLAB supports multidimensional matrices. You can create n dimensional matrices by specifying n indices. The usual matrix creation functions, zeros, ones, rand, and randn, accept n indices to create such matrices. For example,

A = zeros(4,4,3)   initializes a $4 \times 4 \times 3$ matrix A, and
B = rand(2,4,5,6)   creates a $2 \times 4 \times 5 \times 6$ random matrix B.

So, how do we think about the other (the third, fourth, etc.) dimensions of the matrix? You can write a 2-D matrix on a page of a notebook. Think of the third dimension as the different pages of the notebook. Then matrix A $(=$ zeros(4,4,3)$)$ occupies three pages, each page having a 4×4 matrix of zeros. Now, if we want to change the entry in the fourth row and fourth column of the matrix on page 2 to a number 6, we type A(4,4,2)=6. Thus, the usual rules of indexing apply; we only have to think which matrix in the stack we are accessing. Now, once we have moved out of the page (a 2-D matrix), we have no limitation—we can have several notebooks (fourth dimension), several bookcases full of such notebooks (fifth dimension), several rooms full of such bookcases (sixth dimension), and so forth. The multidimensional indexing makes it easy to access elements in any direction you wish—you can get the first element of the matrices on each page, from each notebook, in each bookcase, from all rooms, with C(1,1,:,:,:,:). The usual indexing rules apply in each dimension; therefore, you can access submatrices with index ranges just as you would for a 2-D matrix.

When you operate on multidimensional matrices, you have to be careful. All linear algebra functions will work only on 2-D matrices. You cannot multiply two matrices if they are 3-D or higher dimensional matrices. Matrix multiplication is not defined for such matrices. All array operations (element-by-element operations) are, however, valid for any dimensional matrix. Thus, 5*A, sin(A), and log(A) are all valid commands for matrix A of any dimension. Similarly, if A and B are two matrices of the same dimension, then A+B or A-B is valid irrespective of the dimensionality of A and B.

4.4.2   Structures

A *structure* is a data construct with several named *fields*. Different fields can have different types of data, but a single field must contain data of the same type.

A structure is like a record. One record (structure) can contain information (data) about various things under different heads (fields). For example, you could maintain a record book with one page devoted to each of your relatives. You could list information about them under the headings *relationship, address, name of children, date of birth of all children*, etc. Although the headings are the same in each record, the information contained under a heading could vary in length from record to record. To implement this record-keeping in MATLAB, what you have is a structure. But, what's more, you are not limited to a serial array of record pages (your record book), so you can have a multidimensional structure.

Let us now look at an example of a structure. We are going to make a structure called `FallSem` with fields `course`, `prof`, and `score`. We wish to record the names of courses, names of corresponding professors, and your performance in tests in those courses. Here is one way to do this:

```
FallSem.course = 'cs101';
FallSem.prof = 'Turing';
FallSem.score = [80 75 95];
```

Thus, *fields* are indicated by adding their names after the name of the structure, with a dot separating the two names. The fields are assigned values, just as any other variable in MATLAB. In the preceding example structure, `FallSem`, the fields `course` and `prof` contain character strings, and `score` contains a vector of numbers. Now, how do we generate the record for the next course? Well, structures as well as their fields can be multidimensional. Therefore, we can generate the next record in a few different ways:

Multiple records in a structure array: We can make the structure `FallSem` to be an array (in our example a vector will suffice) and then store a complete record as one element of the array:

```
FallSem(2).course = 'phy200';    FallSem(3).course = 'math211';
FallSem(2).prof = 'Fiegenbaum';  FallSem(3).prof = 'Ramanujan';
FallSem(2).score = [72 75 78];   FallSem(3).score = [85 35 66];
```

Thus, we have created a structure array `FallSem` of size 1×3. Each element of `FallSem` can be accessed just as you access an element of an usual array—`FallSem(2)` or `FallSem(1)`, etc. By typing `FallSem(1)`, you get the values of all the fields, along with the field names. You can also access individual fields, for example, `FallSem(1).prof` or `FallSem(1).score(3)`. See Fig. 4.1 for example output.

In a structure array, each element must have the same number of fields. Each field, however, can have data of different sizes. Thus, `FallSem(1).score` can be a three-element-long row vector, whereas `FallSem(2).score` can be a five-element-long column vector.

```
>> FallSem.course = 'cs101';
>> FallSem.prof = 'Turing';
   FallSem.score = [80 75 95];
```
Create a structure `FallSem` with three fields: `course`, `prof`, and `score`.

```
>> FallSem

   FallSem =
       course: 'cs101'
         prof: 'Turing'
        score: [80 75 95]
```
When queried, MATLAB shows the entire structure.

```
>> FallSem(2).course = 'phy200';    FallSem(3).course = 'math211';
>> FallSem(2).prof = 'Fiegenbaum';  FallSem(3).prof = 'Ramanujan';
>> FallSem(2).score = [72 75 78];   FallSem(3).score = [85 35 66];
```

```
>> FallSem

   FallSem =
   1x3 struct array with fields:
       course
       prof
       score
```
After adding two more records, `FallSem` becomes an array, and when queried, MATLAB now gives structural information about the structure.

```
>> FallSem(2).course

   ans =
   phy200
```
Use array index on the structure to access its elements.

```
>> FallSem(3).score(1)

   ans =
       85
```
You can use index notation for the structure as well as its fields.

```
>> FallSem.score

   ans =
       80    75    95
   ans =
       72    75    78
   ans =
       85    35    66
```
When no index is specified for the structure, MATLAB displays the field values of all records.

```
>> for k=1:3,
       all_scores(k,:) = FallSem(k).score;
   end
```
Use a loop to assign values from a field of several records.

Figure 4.1: A tutorial lesson on structures.

Multiple records in field arrays: For the example we have chosen, we could store multiple records in a single structure (i.e., keep `FallSem` as a 1×1 structure) by making the fields to be arrays of appropriate sizes to accommodate the records:

```
FallSem.course = char('cs101','phy200','math211');
FallSem.prof = char('Turing','Fiegenbaum','Ramanujan');
FallSem.score = [80 75 95; 72 75 78; 85 35 66];
```

In this example, the function `char` is used to create separate rows of character strings from the input variables. Here, `FallSem` is a 1×1 structure, but the field `course` is a 3×7 character array, `prof` is a 3×10 character array, and `score` is a 3×3 array of numbers.

This example works out nicely because we could create a column vector for course names, a column vector for professors' names, and a matrix for scores where each row corresponds to a particular course. What if we had the third record as a matrix for each course? We could still store the record in both ways mentioned. Even though the first method of creating a structure array seems to be the easiest, we could also use the second method and have the third field `score` be a 3-D matrix.

Creating structures

In the preceding examples, we have already seen how to create structures by direct assignment. Just as you create a matrix by typing its name and assigning values to it—`A=[1 2; 3 4];`—you can create a structure by typing its name along with a field and assigning values to the field. This is what we did in the examples. The other way of creating a structure is with the `struct` function. The general syntax of `struct` is

    str_name = struct('fieldname1', 'value1', 'fieldname2', 'value2',···)

Thus, the structure `FallSem` created earlier could also be created using the `struct` function as follows:

As a single structure:

```
FallSem = struct('course',char('cs101','phy200','math211'),...
          'prof', char('Turing','Fiegenbaum','Ramanujan'),...
          'score',[80 75 95; 72 75 78; 85 35 66]);
```

As a structure array:

```
Fall_Sem = [struct('course','cs101','prof','Turing',...
                  'score',[80 75 95]);
            struct('course','phy200','prof','Fiegenbaum',...
                  'score',[72 75 78]);
            struct('course','math211','prof','Ramanujan',...
                  'score',[85 35 66])];
```

Note that this construction creates a 3×1 (a column vector) structure array `Fall_Sem`.

Manipulating structures

Manipulation of structures is similar to the manipulation of general arrays—access elements of the structure by proper indexing and manipulate their values. There is, however, a major difference; you cannot assign all values of a field *across* a structure array to a variable with colon range specifier. Thus, if `Fall_Sem` is a 3×1 structure array, then

`Fall_Sem(1).score(2)`	is valid and gives the second element of `score` from the first record of `Fall_Sem`,
`Fall_Sem(1).score(:)`	gives all elements of `score` from `Fall_Sem(1)`, same as `Fall_Sem(1).score`, and
`a = Fall_Sem(:).score`	is invalid, does not *assign* `score` from all records to A, though the command `Fall_Sem(:).score` or (`Fall_Sem.score`) displays scores from all records.

So, although you can *see* (on the screen) field values across multiple records with `Fall_Sem(:).score`, you have to use a loop construction to *assign* the values to a variable:

```
for k=1:3, all_scores(k,:)  = Fall_Sem(k).score; end
```

The assignment cannot be done directly with the colon operator because the values of fields from several records are treated as separate entities. The field contents are also allowed to be of different sizes. Therefore, although you can use a `for` loop for assignment, you have to take extra care to ensure that the assignment makes sense. For example, in the earlier `for`, if `Fall_Sem(2).score` has only two test scores, then the assignment will produce an error.

It is clear from the previous examples that you can use indices for structures as well as fields for accessing information. Here, we have used only character arrays and number arrays in the fields. You can, however, also have structures inside structures, but the level of indexing becomes quite involved and requires extra care if you have nested structures.

There are also several functions provided to aid manipulation of structures—`fieldnames`, `setfield`, `getfield`, `rmfield`, `isfield`, etc. The names of most of these functions are suggestive of what they do. To get the correct syntax, please see the on-line help on these functions.

4.4.3 Cells

A cell is the most versatile data object in MATLAB. It can contain any type of data—an array of numbers, strings, structures, or cells. An array of cells is called a *cell array.*

You can think of a cell array as an array of data containers. Imagine putting nine empty boxes on the floor in three rows and three columns. Let us call this arrangement a 3×3 cell, `my_riches`. Now, let us put clothes in one box, shoes in the second box, books in the third box, a computer in the fourth box, all greenbacks and coins in the fifth box, and so on. Each box is a container just like any other box; however, the contents of each are different. Thus, we have an arrangement that can accommodate all kinds of things in the containers and still look superficially similar. Now replace the boxes with MATLAB data containers, `cells`, and replace the shoes with a *matrix*, the clothes with a `string array`, the books with a `structure`, the computer with another `cell` (many little boxes inside the big box), and so on, and what you have got, at the topmost level, is a MATLAB data object called *a cell*.

Let us first create a cell, put some data in it, and then discuss various aspects of cells using this cell as an example:

```
C = cell(2,2);              % create a 2 by 2 cell
                            % cell(2) will do the same
C{1,1} = rand(3);           % put a 3x3 random matrix in the 1st box
C{1,2} = char('john','raj'); % put a string array in the 2nd box
C{2,1} = Fall_Sem;          % put a structure in the 3rd box
C{2,2} = cell(3,3);         % put a 3x3 cell in the 4th box
```

In this example, creating a cell superficially looks like creating an ordinary array, but there are some glaring differences. First, the contents are as varied as we want them to be. Second, there are curly braces, instead of parentheses, on the left side of the assignment. So, what are those curly braces for? Could we use parentheses instead?

A cell is different from an ordinary (number) array in that it distinguishes between the (data) containers and the contents, and it allows access to both, separately. When you treat a cell as an array of containers, without paying any attention to the contents, the cell behaves just as an array and you can access a container with the familiar indexing syntax, `C(i,j)`. What you get is the container located at the ith row and jth column. The container will carry a label that will tell you whether the contents are `double`, `char`, `struct`, or `cell`, and of what dimension. If you want to access the contents of a container, you have to use the special *cell-content-indexing*—indices enclosed within curly braces. Thus, to see the 3×3 random matrix in $C(1,1)$, you type `C{1,1}`. See Fig. 4.2 for some examples.

Creating cells

We have already discussed how to create cells using the `cell` function. We can also create cells directly:

```
C = {rand(3)  char('john','raj');  Fall_Sem  cell(3,3)};
```

This example illustrates that the curly braces are to a cell what square brackets are to an ordinary array when used on the right side of an assignment statement.

```
>> Fall_Sem =
     [struct('course','cs101','prof','Turing','score',[80 75 95]);
      struct('course','phy200','prof','Fiegenbaum','score',[72 75 78]);
      struct('course','math211','prof','Ramanujan','score',[85 35 66])];
```

Create a 3 × 1 structure `Fall_Sem` with three fields.

```
>> C = cell(2,2);
>> C{1,1} = rand(3);
>> C{1,2} = char('john','raj');
>> C{2,1} = Fall_Sem;
>> C{2,2} = cell(3,3);
```

Create a 2 × 2 cell *C* and put different types of data in the individual containers of *C*.

```
>> C

  C =

      [3x3 double]    [2x4 char]

      [3x1 struct]    {3x3 cell}
```

When queried, MATLAB shows what type of contents the containers of *C* have.

```
>> C{1,2}

  ans =
  john
  raj
```

Use *content indexing*, {i,j}, on the cell to access the contents of the container at location (i, j).

```
>> C(1,2)

  ans =
      [2x4 char]
```

Use array index notation to access a container (but not its contents).

```
>> C{1,2}(1,:)

  ans =
  john
```

You can use multiple index notation to access a particular data in a particular container.

```
>> C{2,1}(3).prof

  ans =
  Ramanujan
```

You can also combine the structure notation with cell indices.

```
>> C{2,1}(3).score(3)

  ans =
      66
```

This command locates `Fall_Sem(3).score(3)` in *C*.

Figure 4.2: A tutorial lesson on cells.

Manipulating cells

Once you get used to curly braces versus parentheses on both sides of an assignment statement, manipulating cells is just as easy as manipulating an ordinary array. However, there is much more fun to be had with cell indexing, the type of indexing fun you have not had before. Create the cell C given in the preceding example along with the structure Fall_Sem given on page 124, Section 4.4.2. Now try the following commands and see what they do.

```
C(1,1) = {[1 2; 3 4]};
C{1,2}(1,:)
C{2,1}(1).score(3) = 100;
C{2,1}(2).prof
C{2,2}{1,1} = eye(2);
C{2,2}{1,1}(1,1) = 5;
```

There are several functions available for cell manipulation. Some of these functions are cellstr, iscellstr, cell2struct, struct2cell, iscell, num2cell, etc. See the on-line help on these functions for correct syntax and usage.

Two cell functions deserve special mention. These functions are

celldisp   displays the contents of a cell on the screen,
                works recursively on cells inside cells, and

cellplot   plots the cell array schematically.

The result of cellplot(C) is shown in Fig. 4.3 for cell C created in Fig. 4.2. The contents of each cell are shown schematically as arrays of appropriate sizes. The nonempty array elements are shown shaded.

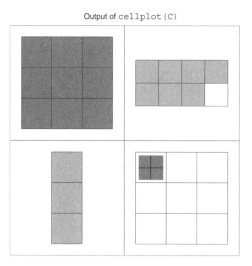

Output of cellplot(C)

Figure 4.3: Cells can be displayed schematically with the function cellplot.

4.5   Publishing Reports

You can create a fancy, formatted report in HTML, XML, LaTeX, MS Word, or MS PowerPoint (MS stuff only on PCs) using MATLAB's built-in *publisher*. For a quick introduction, see Lesson 11 in Chapter 2. Here we discuss, step by step, how to generate a report using the MATLAB publisher.

1. **Create a script file:** First, create a script file containing all commands required to do the computation or produce the desired result. Include comments where necessary to clarify the commands. Let us take, for example, the following script file.

   ```
   % Plot a spiral given by r = e^(theta/10) for theta from 0 to 10*pi
   % Create vectors theta and r
   theta = linspace(0,10*pi,200); % 200 points between 0 & 10*pi
       r = exp(-theta/10);        % compute r
   % plot the spiral using polar plot
   polar(theta,r)
   ```

2. **Format the script as a cell script:** A cell[4] is basically a set of related commands that form one *unit* of computation, or a chunk of descriptive text that forms a unit. A cell could contain one command, many commands, or even no commands. The beginning of a cell is marked by the double percent character (%%). A cell continues until the beginning of another cell is encountered. The text in the line beginning with the %% character is used as the section heading by the publisher. All such headings in a script are used by the publisher to create a hyperlinked table of contents in the report. It is generally a good idea to have an overall title for the report, generated by a cell containing only the title, no commands, at the top of the script file.

   ```
   %% Publishing Reports - A Simple Example
   %% Description
   % Plot a spiral given by r = e^(theta/10), 0 <=theta <= 10*pi
   %
   %% Create vectors theta and r
   theta = linspace(0,10*pi,200); % 200 points between 0 & 10*pi
       r = exp(-theta/10);        % compute r
   %% Plot the spiral using polar plot
   polar(theta,r)
   ```

3. **Format mathematical equations:** Any mathematical equation or expression describing the computation can be formatted using the embedded LaTeX to make it appear like an equation rather than the text shown in the script file. We would much rather have the spiral equation appear as $r = e^{\frac{\theta}{10}}, 0 \le \theta \le 10\pi$ than how it appears now. So, let us format the equation using **Cell → Insert Text Markup → TeX Equation** as follows:

   ```
   %%
   % $$r(\theta) = e^{-\frac{\theta}{10}}, \quad 0\le\theta\le 10\pi$$
   ```

---

[4]This cell is not to be confused with the *cell* data type. Here, it refers to its usual meaning of a unit or a set. It is unfortunate that MATLAB uses the same term in these two contexts.

Note that the mathematical expression is enclosed within the double dollar signs and starts with a % sign. The equation must be preceded by a cell marker (%%). The TeX Equation cell introduced earlier results in the following equation (\quad introduces some blank space):

$$r(\theta) = e^{-\frac{\theta}{10}}, \quad 0 \le \theta \le 10\pi$$

If you do not know LaTeX, see Table 4.1, which shows an introductory list of LaTeX commands used in mathematical expressions. These LaTeX commands are also useful for labeling the axes in figures (e.g., xlabel{'x_{n+1}'}, ylabel{'\log{\dot{x}_n}}).

4. **Bells and whistles:** It is also possible to format the text with boldface letters, monospaced text (to mark MATLAB commands in comments), and bulleted list. You can use these text mark-ups with Cell → Insert Text Markup selections (see Fig. 4.4). For example, boldface text is created by enclosing the text with * on both sides of the text, and monospaced text is created by enclosing it within vertical bars (|). So, with just a smattering of these formats, here is our script file ready to be published.

```
%% Publishing Reports - A Simple Example
%% Description
% Plot a spiral given by
%
% $$r(\theta) = e^{-\frac{\theta}{10}}, \quad 0\le\theta\le 10\pi$$
% using the following steps:
%
% * create data vectors theta and r
% * plot the spiral using function |polar|

%% Create vectors theta and r
theta = linspace(0,10*pi,200); % 200 points between 0 & 10*pi
    r = exp(-theta/10);        % compute r
%% Plot the spiral using polar plot
polar(theta,r)
```

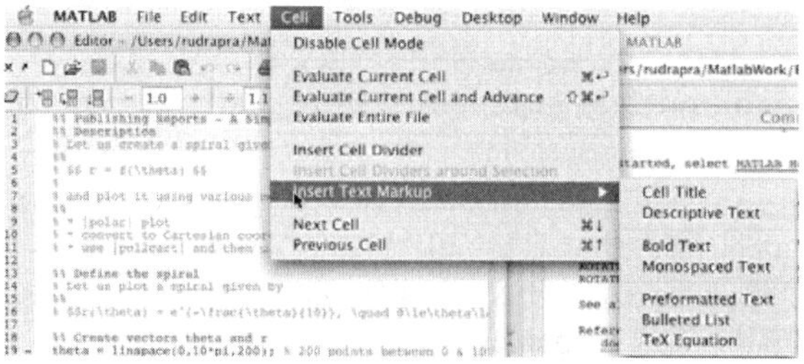

Figure 4.4: The Cell menu of the MATLAB editor contains several formatting tools for the cell script.

5. **Publish:** Write and save the cell script given above in, say, a spiralplot.m file. Open this file in the MATLAB editor and select File → Publish spiralplot.m. MATLAB should create a directory named html, if it does not already exist, and create a document with the name spiralplot.html, along with some other helper files. MATLAB will also automatically open this file for you to see. The file should be nicely formatted and include the figure generated by the `polar` command.

The size of a figure appearing in the report is controlled by setting the image height or width in Publish Configuration for *filename* → Edit Publish Configurations for *filename* dialog box.

You can use the same steps to publish the report in other file formats by selecting the appropriate format from the **Output settings** in the **Edit Publish Configuration** dialog box. Alternatively, you can use the `publish('ScriptName','format')` command in the command window, e.g., `publish('spiralplot','html');`.

Another possibility, if you are a PC user and love MS Word, is to create a fancy **Notebook**—a Word document with embedded MATLAB commands that can be executed from within Word and the output, including graphics, can be included in the document. The created Word document is called an M-book. You can learn how to create an M-book and use it by going to helpdesk → MATLAB → Publishing results → Notebook for publishing to Word. We, however, recommend the publisher discussed previously.

Description	Command	Example	Output
Greek symbols	\name	\alpha,\theta,\beta,\pi \Theta,\Gamma,\Sigma,\phi	$\alpha,\ \beta,\ \theta,\ \pi$ $\Theta,\ \Gamma,\ \Sigma,\ \phi$
Functions	\name	\sin,\arccos \log,\exp	sin, arccos log, exp
Super- and subscript	^{ } and _{ }	\sin^{2}x_{n}+x_{n-1}^{k}	$\sin^2 x_n + x_{n-1}^k$
Fraction	\frac{num}{denom}	\frac{x+|y|}{n^2-\sinh x}	$\frac{x+\|y\|}{n^2-\sinh x}$
Square root	\sqrt{ }	\sqrt{3x^{4} -1}	$\sqrt{3x^4-1}$
Sum and integral	\sum, \int	\sum_{n=0}^{k}\frac{x^{n}}{n!}, \int_{0}^{\infty} e^{-x^2} dx	$\sum_{n=0}^{k}\frac{x^n}{n!}$, $\int_0^\infty e^{-x^2}\,dx$
Derivatives	\frac{d}{dx} \dot{ },\ddot{ } ', ''	\frac{dy}{dx} \dot{x},\ddot{x} y', y''	$\frac{dy}{dx}$ $\dot{x},\ \ddot{x}$ $y',\ y''$

Table 4.1: Table of some LaTeX commands

EXERCISES

1. **A script file to compute sine series:** Write a script file named sineseries.m that computes the value of $\sin(x)$ at a given x using n terms of the series expansion of the sine function:

$$\sin(x) = x - \frac{x^3}{3!} + \frac{x^5}{5!} - \cdots = \sum_{k=1}^{n} (-1)^{k-1} \frac{x^{2k-1}}{(2k-1)!}$$

Follow the steps given here.

- First, query MATLAB to see if the name *sineseries* is already taken by some variable or function with the command exist('sineseries'). What does the MATLAB response mean? [Hint, see on-line help on exist.]
- Include the following line as the header (H1 line) of your script file.
  %SINESERIES: computes sin(x) from series expansion
  Now code the formula so that it computes the sum of the series for a given scalar x and a given integer n.
- Save the file. Type help sineseries to see if MATLAB can access your script file. Now, compute $\sin(\pi/6)$ with $n = 1, 5, 10$, and 20. Compare the results. Do the same for some other x of your choice.

2. **A function file to compute sine series:** Take the script file written in Exercise 1 and convert it into a function file using the following steps.

- Name the function sine_series and modify the H1 line appropriately.
- Let x and n be the input to your function and y (the sum) be the output.
- Save the function and execute it to see that it works and gives the same output as the script file in Exercise 1.
- Modify the function to include more on-line help on how to run the function.
- Modify the function so that it can accept a vector x and give an appropriate y.
- Modify the function to include a check on the input n. The function should proceed only if $n > 0$ is an integer, otherwise it should display an error message.
- Provide for an optional output err, which gives the % error in y when compared to sin(x). [Hint: use conditional statement on nargout for optional output.]
- Modify the function so that it takes a default value of $n = 10$ if the user does not specify n. [Hint: use nargin.]
- Execute the function to check all features you have added.

3. **A function as an input to another function:** There are several ways in which a function can be passed in the input argument list of another function. The function to be used in the input list can be written as an inline function (see Section 3.5.1, page 83) or it can be coded in a function file. How the function is passed in the input list depends on how it is coded.
   Code the function $y(x) = \frac{\sin(x)}{x}$ as an inline function sinc and in a function file called sincfun.m. You will use this function in the input list of ezplot (see Section 3.8, page 93) in various ways in the following instructions.

- Use the inline function sinc in the input list of ezplot to plot the function over the default domain.

- Use the function `sincfun` as a string in the input list of `ezplot` to plot the function over the default domain.
- Create a function handle for `sincfun` and use the handle in the input list of `ezplot` to plot the function over the default domain.

4. **Write subfunctions:** In this exercise you will write a simple function to find $\sin(x)$ and/or $\cos(x)$ from series expansion using two subfunctions—`sine_series` and `cosine_series`.

- Write a function named `cosine_series` to evaluate the cosine series (follow Exercise 2)

$$\cos(x) = 1 - \frac{x^2}{2!} + \frac{x^4}{4!} - \cdots = \sum_{k=1}^{n} (-1)^{k-1} \frac{x^{2(k-1)}}{2(k-1)!}$$

Execute and test the function to make sure it works correctly.
- Write a new function named `trigseries` as follows.
  - The input list of the function should include x, n, and another string variable *options*. The user can specify '`sin`', '`cos`', or '`both`' for *options*. The default should be '`both`'. The output list should include y and *err*, as discussed in Exercise 2.
  - The function `trigseries` should call `sine_series` and `cosine_series` as appropriate to compute y and *err*, depending on what the user specifies in *options*. Implement this call sequence with `switch` using `options` as the switch.
  - Program the output y and *err* to be two column arrays if the user asks for '`both`' in *options* or as the default.
- Copy and paste the functions `sine_series` and `cosine_series` as subfunctions below the function `trigseries` in the same file.
- Delete the original files sine_series2.m and cosine_series2.m or rename them something else. Test `trigseries` with various input and *options*. Make sure it works correctly. Now execute `trigseries` with only one input x=[0 pi/6 pi/4 pi/3 pi/2]. Do you get reasonable answers?

5. **Profile a function:** Profile the function `trigseries`, developed in Exercise 4, taking a vector x of 100 equally spaced points between 0 and π as the only input.

6. **Recursion:** Write a function to compute $n!$ using recursion. (Note that this is not the most efficient way to compute $n!$. However, it is conceptually a recursive calculation that is easy to implement and test recursion in MATLAB.)

5. *Applications*

5.1 Linear Algebra

5.1.1 Solving a linear system

Solving a set of linear algebraic equations is easy in MATLAB. It is, perhaps, also the most used computation in science and engineering. Solving a system of equations on a computer is now as basic a task as doing scientific calculations on a calculator. Let us see how MATLAB makes this task easy and pleasant.

We will solve a set of given linear algebraic equations. To solve these equations, no prior knowledge of matrix methods is required. The first two steps outlined below are really basic for most people who know a little bit of linear algebra. We will consider the following set of equations for our example:

$$5x = 3y - 2z + 10$$
$$8y + 4z = 3x + 20$$
$$2x + 4y - 9z = 9$$

Step 1: Rearrange equations: Write each equation with all unknown quantities on the left-hand side and all known quantities on the right-hand side. Thus, for the preceding equations, rearrange them such that all terms involving x, y, and z are on the left side of the equal sign:

$$
\begin{aligned}
5x - 3y + 2z &= 10 \\
-3x + 8y + 4z &= 20 \\
2x + 4y - 9z &= 9
\end{aligned}
\tag{5.1}
$$

Step 2: Write the equations in matrix form: To write the equation in the matrix form $[\mathbf{A}]\{\mathbf{x}\} = \{\mathbf{b}\}$ where $\{\mathbf{x}\}$ is the vector of unknowns, you have to arrange the unknowns in vector $\mathbf{x}$, the coefficients of the unknowns in matrix

A, and the constants on the right-hand side of the equations in vector **b**. In this particular example, the unknown vector is

$$\mathbf{x} = \begin{bmatrix} x \\ y \\ z \end{bmatrix},$$

the coefficient matrix is

$$\mathbf{A} = \begin{bmatrix} 5 & -3 & 2 \\ -3 & 8 & 4 \\ 2 & 4 & -9 \end{bmatrix},$$

and the known constant vector is

$$\mathbf{b} = \begin{bmatrix} 10 \\ 20 \\ 9 \end{bmatrix}.$$

Note that the columns of **A** are simply the coefficients of each unknown from all three equations. Now you are ready to solve this system in MATLAB.

Step 3: Solve the matrix equation in MATLAB: Enter the matrix **A** and vector **b**, and solve for vector **x** with x=A\b (note that the \ is different from the division /):

For on-line help type:
`help slash`

```
>> A = [5 -3 2; -3 8 4; 2 4 -9];   % Enter matrix A
>> b = [10; 20; 9];                % Enter column vector b
>> x = A\b                         % Solve for x
x =
    3.4442
    3.1982
    1.1868

>> c = A*x                         % check the solution
c =
   10.0000
   20.0000
    9.0000
```

The backslash (\) or the left division is used to solve a linear system of equations $[\mathbf{A}]\{\mathbf{x}\} = \{\mathbf{b}\}$. For more information, type: `help slash`.

You can also use a powerful function, `linsolve`, to solve linear systems if your matrix **A** has exploitable structure (e.g., triangular, positive definite).

5.1.2 Gaussian elimination

In introductory linear algebra courses, we learn to solve a system of linear algebraic equations by *Gaussian elimination*. This technique requires forming a rectangular matrix that contains both the coefficient matrix **A** and the known vector **b** in an *augmented matrix*. Gauss-Jordan reduction procedure is then used to transform the augmented matrix to the so-called *row reduced echelon form*. MATLAB has a built-in function, `rref`, that does precisely this reduction, i.e., transforms the matrix to its `r`ow `r`educed `e`chelon `f`orm. For example, consider eqn. (5.1) with the coefficient matrix **A** and the known vector **b** created previously. To solve the equations with `rref`, type the following commands:

```
C = [A  b];        % form the augmented matrix
Cr = rref(C);      % row reduce the augmented matrix
```

The last column of `Cr` is the solution `x`. You may like to use `rref` for checking homework solutions; it is not very useful for anything else.

Eigenvalues: (~ natural frequencies)

$$\begin{bmatrix} 5-\lambda, & -3, & 2 \\ -3, & 8-\lambda, & 4 \\ 4, & 2, & -9-\lambda \end{bmatrix} = 0$$

$= \lambda_1, \lambda_2, \lambda_3$

Eigenvectors: (~ amplitude)

5.1.3 Finding eigenvalues and eigenvectors

The omnipresent eigenvalue problem in scientific computation shows up as

$$\mathbf{A}\ \mathbf{v} = \lambda \mathbf{v} \tag{5.2}$$

λ is a scalar. The problem is to find λ and $\mathbf{v}$ for a given $\mathbf{A}$ so that eqn. (5.2) is satisfied. By hand, the solution is usually obtained by first solving for the n eigenvalues from the determinant equation $|\mathbf{A} - \lambda\mathbf{I}| = 0$, and then solving for the n eigenvectors by substituting the corresponding eigenvalues in eqn. (5.2), one at a time. On pencil and paper, the computation requires a few pages even for a 3×3 matrix,[1] but it is a 1.2-inch long command in MATLAB (excluding entering the matrix)! Here is an example:

Step 1: Enter matrix **A** and type `[V,D]=eig(A)`. (See Fig. 5.1.)

V= eigen vector

D = eigen value

```
>> A = [5 -3  2; -3  8  4; 4  2  -9];
>> [V,D]  =  eig(A)

V =

    0.1725    0.8706   -0.5375
    0.2382    0.3774    0.8429
   -0.9558    0.3156   -0.0247


D =

  -10.2206         0         0
         0    4.4246         0
         0         0    9.7960
```

Here **V** is a matrix containing the eigenvectors of **A** as its columns. For example, the first column of **V** is the first eigenvector of **A**.

D is a matrix that contains the eigenvalues of **A** on its diagonal.

Figure 5.1: Finding eigenvalues and eigenvectors of a matrix.

[1]If **A** is bigger than 4×4, you have to be insane to try to solve it by hand; for a 4×4 matrix, you are either borderline insane or you live in a civilization without computers.

Step 2: Extract what you need: In the output list, **V** is an $n \times n$ matrix whose columns are eigenvectors and **D** is an $n \times n$ diagonal matrix that has the eigenvalues of **A** on its diagonal. The function `eig` can also be used with one output argument, e.g., `lams=eig(A)`, in which case the function gives only the eigenvalues in the vector `lams`.

After computing the eigenvalues and eigenvectors, you could check a few things if you wish: Are the eigenvalues and eigenvectors ordered in the output? Check by substituting in eqn. (5.2). For example, let us check the second eigenvalue and second eigenvector and see if, indeed, they satisfy $Av = \lambda v$.

```
>> v2 = V(:,2);    % extract the second column from V
>> lam2 = D(2,2);  % extract the second eigenvalue from D
>> A*v2 - lam2*v2  % check the difference between A*v and lambda*v
```

Here, we have used two commands for clarity. You could, of course, combine them into one: `A*V(:,2)-D(2,2)*V(:,2)`. Note that the last command may not result into zero vector, as you expect, but it will have very small numbers.

5.1.4   Matrix factorizations

MATLAB provides built-in functions for several matrix factorizations (decompositions):

1. **LU factorization:** The name of the built-in function is `lu`. To get the LU factorization of a square matrix **A**, type the command

   ```
   [L,U] = lu(A);
   ```

   MATLAB returns a lower triangular matrix **L** and an upper triangular matrix **U** such that L*U=A. It is also possible to get the permutation matrix as an output (see the on-line help on `lu`).

2. **QR factorization:** The name of the built-in function is `qr`. Typing the command

   ```
   [Q,R] = qr(A);
   ```

   returns an orthogonal matrix **Q** and an upper triangular matrix **R** such that Q*R=A. For more information, see the on-line help.

3. **Cholesky factorization:** If you have a positive definite matrix **A**, you can factorize the matrix with the built-in function `chol`. The command

   ```
   R = chol(A);
   ```

   produces an upper triangular matrix **R** such that R'*R=A.

4. **Singular value decomposition (svd):** The name of the built-in function is `svd`. If you type

   ```
   [U,D,V] = svd(A);
   ```

   MATLAB returns two orthogonal matrices, **U** and **V**, and a diagonal matrix **D**, with the *singular values* of **A** as the diagonal entries, such that U*D*V=A.

If you know which factorization to use, then all you need to know here is the syntax of the corresponding function. If you do not know where and how these factorizations are used, then this is not the right place to learn it; look into your favorite books on linear algebra.[2]

5.1.5 Advanced topics

MATLAB's main strength is its awesome suite of linear algebra functions. There are hundreds of functions that aid in finding solutions to linear algebra problems, from very basic problems to very advanced ones. For example, there are more than 10 eigenvalue-related functions. Type `lookfor eigenvalue` to see a list of these functions. Among these functions, `eigs` deserves special mention. This function was added in MATLAB 5 to help find just a few eigenvalues and eigenvectors, a requirement that frequently arises in large-scale problems.

For on-line help type: `help sparfun`

There is a separate suite of functions for sparse matrices and computations with these matrices. These functions include

Category	Example functions
Elementary matrix functions	speye, sprand, spdiags
Full to sparse conversion	sparse, full, spconvert
Utility functions	nnz, nzmax, spalloc, spy
Reordering algorithm	colamd, colperm, symamd, symrcm
Linear algebra	eigs, svds, luinc, cholinc
Linear equations	pcg, bicg, cgs, qmr

Please see the on-line help with `help sparfun` for a list of these functions and their brief descriptions. Detailed on-line help is available on each function.

MATLAB also provides several functions for operations on graphs or trees, such as `treelayout`, `treeplot`, `etree`, `etreeplot`, `gplot`, etc. These functions are listed under the `sparfun` category.

5.2 Curve Fitting and Interpolation

5.2.1 Polynomial curve fitting on the fly

Curve fitting is a technique of finding an algebraic relationship that "best" (in a *least squares* sense) fits a given set of data. Unfortunately, there is no magical function (in MATLAB or otherwise) that can give you this relationship if you simply supply the data. You have to have an idea of what kind of relationship might exist between the input data (x_i) and the output data (y_i). However, if you do not have a firm idea but you have data that you trust, MATLAB can help you explore the best possible fit. MATLAB includes **Basic Fitting** in its figure window's **Tools** menu that lets you fit a polynomial curve (up to the tenth order) to your data on the fly. It also

---

[2]Some of my favorites: *Linear Algebra and Its Applications* by Strang, Saunders HBJ College Publishers; *Matrix Computations* by Golub and Van Loan, The Johns Hopkins University Press; *Matrix Analysis* by Horn and Johnson, Cambridge University Press.

gives you options of displaying the residual at the data points and computing the norm of the residuals. This can help in comparing different fits and then selecting the one that makes you happy. Let us take an example and go through the steps.

Example 1: Straight-line (linear) fit

Let us say that we have the following data for x and y and we want to get the best linear (straight-line) fit through this data.

x	5	10	20	50	100
y	15	33	53	140	301

Here is all it takes to get the best linear fit, along with the equation of the fitted line.

Step 1: Plot raw data: Enter the data and plot it as a scatter plot using some marker, say, circles.

```
x = [5 10 20 50 100];      % x-data
y = [15 33 53 140 301];    % y-data
plot(x,y,'o')              % plot x vs y using circles
```

Step 2: Use built-in Basic Fitting **to do a linear fit:** Go to your figure window, click on Tools, and select Basic Fitting from the pull-down menu (see Fig. 5.2). A separate window appears with Basic Fitting options.

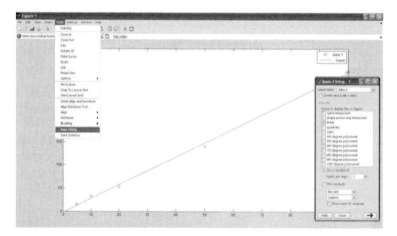

Figure 5.2: First plot the raw data. Then select Basic Fitting from the Tools menu of the figure window. A separate window appears with several Basic Fitting options for polynomial curve fits. Check appropriate boxes in this window to get the desired curve fit and other displays, such as the fitted equation.

Step 3: Fit a linear curve and display the equation: Check the boxes for linear and Show equations from the Basic Fitting window options. The best fitted line as well as its equation appear in the figure window. You are now done. The result is shown in Fig. 5.3.

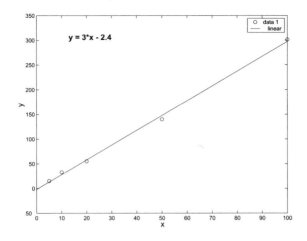

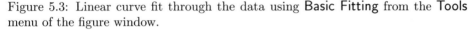

Figure 5.3: Linear curve fit through the data using **Basic Fitting** from the **Tools** menu of the figure window.

Example 2: Comparing different fits

Let us take another example where we try two different fits, quadratic and cubic, for the same data and do a comparison to figure out which one is better. We first create some x and y data.

```
x = 0 : pi/30 : pi/3;        % x-data
y = sin(x) + rand(size(x))/100  % y-data (corrupted sine)
```

Step 1: Plot raw data: We already have the x and y data. So, go ahead and plot the raw data with `plot(x,y,'o')`.

Step 2: Use Basic Fitting to do a quadratic and a cubic fit: Go to your figure window, click on **Tools**, and select **Basic Fitting** from the pull-down menu (as in Example 1). In the **Basic Fitting** window, check **quadratic** and **cubic** boxes. In addition, check the boxes for **Show equations**, **Plot residuals**, and **Show norm of residuals**.

The result is shown in Fig. 5.4. The residual at each data point is plotted in the lower subplot for each fit and the norm of the residual is also shown on the plot. Note that although the two curves are visually almost indistinguishable over the range of the data, the norm of the residuals for the cubic fit is an order of magnitude lower than that of the quadratic fit. Therefore, you may like to chose the cubic fit in this case.

5.2.2   Do it yourself: Curve fitting with polynomial functions

It may be worth understanding how these curve fits work. In MATLAB, it is fairly easy to do polynomial curve fits using built-in polynomial functions and get the desired coefficients. If you would like to understand it, please read on, otherwise you may want to skip this section.

For on-line help type:
`help polyfun`

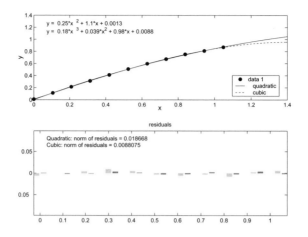

Figure 5.4: A comparison of quadratic and cubic curve fits for some data using Basic Fitting from the Tools menu of the figure window. (Note that if you try this example, you will get slightly different coefficients and residuals from those shown here because y contains *random* noise.)

The simplest relationship between two variables, say x and y, is a linear relationship: $y = mx + c$. For a given set of data points (x_i, y_i), the problem is to find m and c such that $y_i = mx_i + c$ best fits the data. For data points that are not linearly related, you may seek a polynomial relationship

$$y_i = a_k x_i^k + a_{k-1} x_i^{k-1} + \cdots + a_2 x_i^2 + a_1 x_i + a_0$$

or an exponential relationship $y_i = c_1 e^{c_2 x_i}$ or even more complicated relationships involving logarithms, exponentials, and trigonometric functions.

For polynomial curve fits, of any order n, the problem is to find the $(n + 1)$ coefficients $a_n, a_{n-1}, a_{n-2}, \cdots a_1$, and a_0, from the given data of length $(n+1)$ or more. MATLAB provides an easy way—through the built-in functions `polyfit` and `polyval`.

`polyfit` Given two vectors x and y, the command `a = polyfit(x,y,`n`)` fits a polynomial of order n through the data points (x_i, y_i) and returns $(n+1)$ coefficients of the powers of x in the row vector a. The coefficients are arranged in the decreasing order of the powers of x, i.e., `a = [`a_n a_{n-1} $\cdots$ a_1 a_0`]`.

`polyval` Given a data vector x and the coefficients of a polynomial in a row vector a, the command `y=polyval(a,x)` evaluates the polynomial at the data points x_i and generates the values y_i such that

$$y_i = a(1)x_i^n + a(2)x_i^{n-1} + \cdots + a(n)x + a(n+1).$$

Here the length of the vector a is $n + 1$ and, consequently, the order of the evaluated polynomial is n. Thus if a is five elements long, the polynomial to be evaluated is automatically ascertained to be of fourth order.

Both `polyfit` and `polyval` use an optional argument if you need error estimates. To use the optional argument, see the on-line help on these functions.

Example: Straight-line (linear) fit

The following data is obtained from an experiment aimed at measuring the spring constant of a given spring. Different masses m are hung from the spring and the corresponding deflections δ of the spring from its unstretched configuration are measured. From physics, we know that $F = k\delta$ and here $F = mg$. Thus, we can find k from the relationship $k = mg/\delta$. Here, however, we are going to find k by plotting the experimental data, fitting the best straight line (we know that the relationship between δ and F is linear) through the data, and then measuring the slope of the best-fit line.

m(g)	5.00	10.00	20.00	50.00	100.00
δ(mm)	15.5	33.07	53.39	140.24	301.03

Fitting a straight line through the data means we want to find the polynomial coefficients a_1 and a_0 (a first-order polynomial) such that $a_1 x_i + a_0$ gives the "best" estimate of y_i. In steps, we need to do the following.

Step 1: Find the coefficients a_k's:

```
a = polyfit(x,y,1)
```

Step 2: Evaluate y at finer (more closely spaced) x_j's using the fitted polynomial:

```
y_fitted = polyval(a,x_fine)
```

Step 3: Plot and see. Plot the given data as points and fitted data as a line:

```
plot(x,y,'o',x_fine,y_fitted);
```

As an example, the following script file shows all the steps involved in making a straight-line fit through the given data for the spring experiment and finding the spring constant. The resulting plot is shown in Fig. 5.5

```
m=[5 10 20 50 100];              % mass data (g)
d=[15.5 33.07 53.39 140.24 301.03]; % displacement data (mm)
g=9.81;                          % g = 9.81 m/s^2
F=m/1000*g;                      % compute spring force (N)
a=polyfit(d,F,1);                % fit a line (1st order polynomial)
d_fit=0:10:300;                  % make a finer grid on d
F_fit=polyval(a,d_fit);          % evaluate the polynomial at new points
plot(d,F,'o',d_fit,F_fit)        % plot data and the fitted curve
xlabel('Displacement \delta (mm)'),ylabel('Force (N)')
k=a(1);                          % Find the spring constant
text(100,.32,['\leftarrow Spring Constant K = ',num2str(k),' N/mm']);
```

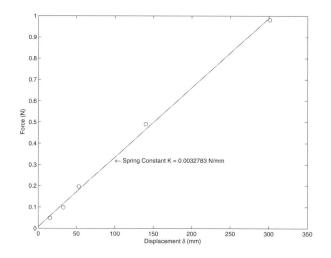

Figure 5.5: A straight-line fit (a polynomial fit of order one) through the spring data.

Comments: Even though you can do a polynomial curve fit of any order, you should have a good reason, possibly based on the physical phenomena behind the data, for trying a particular order. Blindly fitting a higher-order polynomial is often misleading. The most common and trustworthy curve fit, by far, is the straight line. If you expect the data to have exponential relationship, convert the data to a log scale and then do a linear curve fit. The result is a lot more trustworthy than an arbitrary sixth- or tenth-order polynomial fit. As a rule of thumb, polynomial curve fits of order higher than four or five are rarely required.

5.2.3   Least squares curve fitting

The technique of least squares curve fit can easily be implemented in MATLAB, because the technique results in a set of linear equations that need to be solved. Here, we will not discuss the details of the technique itself because of space limitations as well as the book's intent. You may want to consult a book on numerical methods if you do not know the underlying principles.

Most of the curve fits we do are either polynomial curve fits or exponential curve fits (includes power laws, e.g., $y = ax^b$). If we want to fit a polynomial of order n through our data, we can use MATLAB's built-in function `polyfit`, which already does a least squares curve fit. Therefore, no special effort is required. Now what if we want to fit a nonpolynomial function? Two most commonly used functions are:

1. $y = ae^{bx}$

2. $y = cx^d$

We can convert these exponential curve fits into polynomial curve fits (actually a linear one) by taking the log of both sides of the equations, i.e.,

1. $\ln(y) = \ln(a) + bx$ or $\tilde{y} = a_0 + a_1 x$, where $\tilde{y} = \ln(y)$, $a_1 = b$, and $a_0 = \ln(a)$.
2. $\ln(y) = \ln(c) + d\ln(x)$ or $\tilde{y} = a_0 + a_1\tilde{x}$, where $\tilde{y} = \ln(y)$, $a_1 = d$, $a_0 = \ln(c)$ and $\tilde{x} = \ln x$.

Now we can use `polyfit` in both cases with just first-order polynomials to determine the unknown constants. The steps involved are the following.

Step 1: Prepare new data: Prepare new data vectors $\tilde{y}$ and $\tilde{x}$, as appropriate, by taking the log of the original data. For example, to fit a curve of the type $y = ae^{bx}$, create `ybar=log(y)` and leave x as it is; to fit a curve of the type $y = cx^d$, create `ybar=log(y)` and `xbar=log(x)`.

Step 2: Do a linear fit: Use `polyfit` to find the coefficients a_0 and a_1 for a linear curve fit.

Step 3: Plot the curve: From the curve fit coefficients, calculate the values of the original constants (e.g., a, b). Recompute the values of y at the given x's according to the relationship obtained and plot the curve along with the original data.

Here is an example of one such curve fit. The following table shows the time versus pressure variation readings from a vacuum pump. We will fit a curve, $P(t) = P_0 e^{-t/\tau}$, through the data and determine the unknown constants P_0 and τ.

t	0	0.5	1.0	5.0	10.0	20.0
P	760	625	528	85	14	0.16

By taking log of both sides of the relationship, we have

$$\ln(P) = \ln(P_0) - \frac{t}{\tau}$$
$$\text{or,} \quad \tilde{P} = a_1 t + a_0$$

where $\tilde{P} = \ln(P)$, $a_1 = -1/\tau$, and $a_0 = \ln(P_0)$. Thus, we can easily compute P_0 and τ once we have a_1 and a_0. The following script file shows all the steps involved. The results obtained are shown both on the linear scale and on the log scale in Fig. 5.6.

```
% EXPFIT: Exponential curve fit example
% For the following data for (t,p) fit an exponential curve
%        p = p0 * exp(-t/tau).
% The problem is solved by taking log and then using a linear
% fit (1st order polynomial)

% original data
t = [0 0.5 1 5 10 20];
p = [760 625 528 85 14 0.16];

% Prepare new data for linear fit
tbar = t;                       % no change in t is required
pbar = log(p);
```

```
% Fit a 1st order polynomial through (tbar,pbar)
a = polyfit(tbar,pbar,1);    % the output is a = [a1 a0]

% Evaluate constants p0 and tau
p0 = exp(a(2));              % since a(2) = a0 = log(p0)
tau = -1/a(1);              % since a1 = -1/tau

% (a) Plot the new curve and the data on linear scale
tnew = linspace(0,20,20);   % create more refined t
pnew = p0*exp(-tnew/tau);   % evaluate p at new t
plot(t,p,'o',tnew,pnew), grid
xlabel('Time (sec)'), ylabel('Pressure (torr)')

% (b) Plot the new curve and the data on semilog scale
lpnew = exp(polyval(a,tnew));
semilogy(t,p,'o',tnew,lpnew),grid
xlabel('Time (sec)'), ylabel('Pressure (torr)')

% Note: you only need one plot, you can select (a) or (b).
```

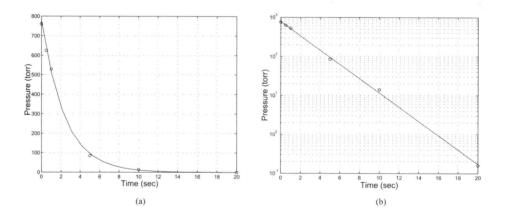

Figure 5.6: Exponential curve fit: (a) linear scale plot, (b) semilog scale plot.

There is yet another way to fit a complicated function through your data in the least squares sense. For example, let us say that you have time (t) and displacement (y) data from a spring-mass system experiment and you think that the data should follow

$$y = a_0 \cos(t) + a_1 t \sin(t)$$

Here the unknowns are only a_0 and a_1. The equation is nonlinear in t but *linear in the parameters* a_0 and a_1. Therefore, you can set up a matrix equation using each data point and solve for the unknown coefficients. The matrix equation is

$$
\begin{bmatrix}
\cos(t_1) & t\sin(t_1) \\
\cos(t_2) & t\sin(t_2) \\
\cos(t_3) & t\sin(t_3) \\
\vdots & \vdots \\
\cos(t_n) & t\sin(t_n)
\end{bmatrix}
\begin{Bmatrix} a_0 \\ a_1 \end{Bmatrix}
=
\begin{Bmatrix} x_1 \\ x_2 \\ x_3 \\ \vdots \\ x_n \end{Bmatrix}.
$$

Now you can solve for the unknowns a_0 and a_1 simply by typing `a=A\x` where **A** is the coefficient matrix and **x** is the vector containing the measured data x_i. Typically, **A** is a rectangular matrix and the matrix equation to be solved is overdetermined (more independent equations than unknowns). This, however, poses no problem, because the backslash operator in MATLAB solves the equation in a least squares sense whenever the matrix is rectangular.

5.2.4 General nonlinear fits

In curve fitting, at times, we need to fit nonlinear equations in which the unknown coefficients appear inside nonlinear functions (as opposed to being linear multipliers to the nonlinear terms). It is not unusual to have double exponential curve fits: $y(x) = C_1 e^{\lambda_1} x + C_2 e^{\lambda_2} x$. Here C_1 and C_2 are linear coefficients but λ_1 and λ_2 are nonlinear coefficients. In MATLAB you can do such curve fits by using the services of `fminsearch`, which helps you find appropriate values of the nonlinear coefficients by minimizing the error arising from a guess for their values. Type `fitdemo` on the command prompt to see a demo of how you can do such curve fits. However, you are not limited to fitting nonlinear functions of only this form. For example, see problem 4 in the Exercises (it is more like an example with step-by-step instructions) to find out how to fit a function such as $x(t) = Ce^{\lambda_1 t}\sin\lambda_2 t$.

5.2.5 Interpolation

Interpolation is the technique of finding a functional relationship between variables such that a given set of discrete values (data points) of the variables satisfy that relationship.[3] Usually, we get a finite set of data points from experiments. When we want to pass a smooth curve through these points or find some intermediate points, we use the technique of interpolation. Interpolation is NOT curve fitting, in that it requires the interpolated curve to pass through all the data points.

In MATLAB, you can interpolate your data using *splines* or *Hermite interpolants* on a fly. All you have to do is plot the raw data and use **spline interpolant** or **hermite interpolant** from the **Basic Fitting** options that you can invoke from the **Tools** menu of your figure window (see Fig. 5.2 on page 140).

In its programming environment, MATLAB provides the following functions to facilitate interpolation:

[handwritten margin note:] Solve for 'y'
$$y = y_1 + \frac{y_2 - y_1}{x_2 - x_1}(x - x_1) =$$

---

[3]I asked some of my colleagues to define interpolation. One gave me this definition, "Reverse engineering an entity of higher dimensionality from information about some entity of lower dimensionality within an identified domain."

`interp1` One-dimensional data interpolation, i.e., given y_i at x_i, finds y_j at desired x_j from $y_j = f(x_j)$. Here f is a continuous function that is found from interpolation. It is called one-dimensional interpolation because y depends on a single variable x. The calling syntax is

$$\text{ynew = interp1(x,y,xnew,\textit{method})}$$

where *method* is an optional argument discussed after the descriptions of `interp2` and `interp3`.

`interp2` Two-dimensional data interpolation, i.e., given z_i at (x_i, y_i), finds z_j at desired (x_j, y_j) from $z = f(x, y)$. The function f is found from interpolation. It is called two-dimensional interpolation because z depends on two variables, x and y.

$$\text{znew = interp2(x,y,z,xnew,ynew,\textit{method}).}$$

`interp3` Three-dimensional analogue of `interp1`, i.e., given v_i at (x_i, y_i, z_i), finds v_j at desired (x_j, y_j, z_j).

$$\text{vnew = interp3(x,y,z,v,xnew,ynew,znew,\textit{method}).}$$

In addition, there is an *n*-dimensional analogue, `interpn`, if you ever need it. In each function, you have an option of specifying a *method* of interpolation. The choices for *method* are *nearest, linear, cubic,* or *spline.* The choice of the method dictates the smoothness of the interpolated data. The default method is *linear.* To specify cubic interpolation instead of linear, for example, in `interp1`, use the syntax

$$\text{ynew = interp1(x,y,xnew,'cubic').}$$

The example at the end of this section shows how to use `interp1`. It also compares results obtained from different interpolation methods.

There are some other important interpolation functions worth mentioning:

`spline` One-dimensional interpolation that uses cubic splines to find y_j at desired x_j, given y_i at x_i. Cubic splines fit separate cubic polynomials between successive data points by matching the slopes as well as the curvature of each segment at the given data points. The calling syntax is

$$\text{ynew = spline(x,y,xnew,\textit{method}).}$$

There are other variants of the calling syntax. It is also possible to get the coefficients of the interpolated cubic polynomial segments that can be used later. See the on-line help.

`interpft` Fast Fourier transform (FFT)–based 1-D data interpolation. This is similar to `interp1` except that the data is interpolated first by taking the Fourier transform of the given data and then calculating the inverse transform using more data points. This interpolation is especially useful for periodic functions (i.e., if values of y are periodic). See the on-line help.

Example:

There are two simple steps involved in interpolation—providing a list (a vector) of points at which you wish to get interpolated data (this list may include points at which data is already available), and executing the appropriate function (e.g., `interp1`) with the desired choice for the method of interpolation. We illustrate these steps through an example on the x and y data given in the following table.

x	0	0.785	1.570	2.356	3.141	3.927	4.712	5.497	6.283
y	0	0.707	1.000	0.707	0.000	−0.707	−1.000	−0.707	−0.000

Step 1: Generate a vector `xi` containing desired points for interpolation.

```
% take equally spaced fifty points.
xi = linspace(0,2*pi,50);
```

Step 2: Generate data yi at xi.

```
% generate yi at xi with cubic interpolation.
yi = interp1(x,y,xi,'cubic');
```

Here, `'cubic'` is the choice for interpolation scheme. The other schemes we could use are **nearest, linear,** and **spline.** The data generated by each scheme is shown in Fig. 5.7, along with the original data. The corresponding curves show the smoothness obtained.

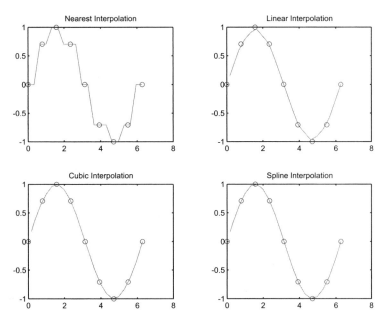

Figure 5.7: 1-D interpolation of data with different schemes: (a) nearest, (b) linear, (c) cubic, and (d) spline.

Caution:
In all interpolation functions, it is required that the input data points in x be monotonic (i.e., either continuously increasing or decreasing).

5.3   Data Analysis and Statistics

Tools menu of
figure window

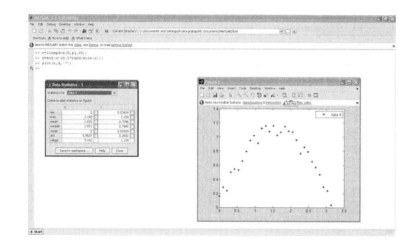

For performing simple data analysis tasks, such as finding mean, median, and standard deviation, MATLAB provides an easy graphical interface that you can activate from the **Tools** menu of the figure window. First, you should plot your data in the form you wish (e.g., scatter plot, line plot). Then, go to the figure window and select **Data Statistics** from the **Tools** pull-down menu. MATLAB shows you the basic statistics of your data in a separate window marked **Data Statistics**. You can show any of the statistical measures on your plot by checking the appropriate box (see Fig. 5.8).

Figure 5.8: Simple data statistics is available to you with the click of a button in MATLAB from the **Tools** menu of the figure window.

However, you are not limited to this simple interface for your statistical needs. Several built-in functions are at your disposal for statistical calculations. These functions are briefly discussed later.

All data analysis functions take both vectors and matrices as arguments. When a vector is given as an argument, it does not matter whether it is a row vector or a column vector. However, when a matrix is used as an argument, the functions operate columnwise on the matrix and output a row vector that contains results of the operation on each column. In the following description of the functions, we will use two arguments in the examples—a row vector x and a matrix A that are given below.

$$x = \begin{bmatrix} 1 & 2 & 3 & 4 & 5 \end{bmatrix}, \qquad A = \begin{bmatrix} 6 & 5 & -2 \\ 7 & 4 & -1 \\ 8 & 3 & 0 \\ 9 & 2 & 1 \\ 10 & 2 & 2 \end{bmatrix}.$$

`mean` gives the arithmetic mean $\bar{x}$ or the average of the data.
Example: `mean(x)` gives 3 while `mean(A)` results in [8 3.2 0].

`median` gives the middle value or the arithmetic mean of the two middle values of the data.
Example: `median(x)` gives 3 while `median(A)` gives [8 3 0].

`std` gives the standard deviation σ based on $n-1$ samples. A flag of value 1 is used as an optional argument (e.g., `std(x,1)`) to get the standard deviation based on n samples.
Example: `std(x)` gives 1.5811, `std(x,1)` gives 1.4142, and `std(A)` gives [1.5811 1.3038 1.5811].

`max` finds the largest value in the data set.
Example: `max(x)` gives 5 and `max(A)` gives [10 5 2].

`min` finds the smallest value in the data set.
Example: `min(x)` gives 1 and `min(A)` gives [6 2 −2].

`sum` computes the sum Σx_i of the data.
Example: `sum(x)` gives 15 while `sum(A)` gives [40 16 0].

`cumsum` computes cumulative sum.
Example: `cumsum(x)` produces [1 3 6 10 15].

`prod` computes the product $\prod x_i$ of all data values.
Example: `prod(x)` gives 120 and `prod(A)` gives [30240 240 0].

`cumprod` computes cumulative product.
Example: `cumprod(x)` produces [1 2 6 24 120].

`sort` sorts the data in ascending (default) or descending order. An optional output argument (e.g., `[y,k]=sort(x)`) gives the indices of the sorted data values.
Example: Let `z=[22 18 35 44 9]`. Then, `sort(z)` results in [9 18 22 35 44], and `[y,j]=sort(z)` gives y=[9 18 22 35 44] and j=[5 2 1 3 4], where elements of j are the indices of the data in z that follow ascending order. Thus, `z=z(j)` will result in a sorted z.

The index vector of the sorted data is typically useful in a situation where you want to sort a few vectors or an entire matrix based on the sorting order of just one vector or column. For example, say you have an $m \times n$ matrix B that you want to sort according to the sorted order of the second column of B. You can do this simply with the commands:

$$[\texttt{z,j}] = \texttt{sort(B(:,2)); Bnew = B(j,:)}$$

For a matrix A, use `sort(A,1)` to sort A by rows and `sort(A,2)` to sort by columns. To sort in descending order, use `sort(A,'descend')`—same as `sort(A,1,'descend')` or `sort(A,2,'descend')`.

Note: You can also get the data in descending order by first sorting the data in the default ascending order and then flipping it upside down with `flipud`. Try `flipud(sort(A))` to get all columns of A arranged in descending order.

`sortrows` sorts the rows of a matrix.

diff computes the difference between the successive data points. For example, y=diff(x) results in a vector y where $y_i = x_{i+1} - x_i$. The resulting vector is one element shorter than the original vector. diff can be used to get approximate numerical derivatives. See the on-line help on diff.

trapz computes the integral of the data (area under the curve defined by the data) using trapezoidal rule.

Example: trapz(x) gives 12 while trapz(A) gives [32 12.5 0].

cumtrapz computes the cumulative integral of the data. Thus, y=cumtrapz(x) results in a vector y where $y_i = \Sigma_{t_1}^{t_i} x_i \Delta t_i$ (here the data x_i is assumed to be taken at t_i, e.g., $x_1 = x(t_1)$, $x_2 = x(t_2)$).

Example: cumtrapz(x) results in [0 1.5 4 7.5 12].

In addition to these most commonly used statistical functions, MATLAB also provides functions for correlation and cross-correlation, covariance, filtering, and convolution. See the on-line help on **datafun**. If you do a lot of statistical analysis, then it may be worth getting the *Statistical Toolbox*.

5.4 Numerical Integration (Quadrature)

For on-line help type:
help funfun

Numerical evaluation of the integral $\int f(x)dx$ is called quadrature. Most often, the integrand $f(x)$ is quite complicated and it may not be possible to carry out the integration analytically. In such cases, we resort to numerical integration. However, we can only evaluate definite integrals, i.e., $\int_a^b f(x)dx$, numerically. There are several methods for numerical integration. Consult your favorite book[4] on numerical methods for a discussion of methods, formulas, and algorithms. MATLAB provides the following built-in functions for numerical integration.

For on-line help type:
help quad

quad integrates a specified function over specified limits, based on adaptive Simpson's rule. The adaptive rule seeks to improve accuracy by adaptively selecting the size of the subintervals (as opposed to keeping it constant) within the limits of integration while evaluating the sums that make up the integral.

quadl (the last letter is an ell, as in QUADL, not 1) integrates a specified function over specified limits, based on adaptive Lobatto quadrature. This one is more accurate than quad but it also uses more function evaluations. It may, however, be more efficient if your integrand is a smooth function.

The general call syntax for both **quad** and **quadl** is as follows:

---

[4]Some of my favorites: *Applied Numerical Analysis* by Gerald and Wheatley, Addison Wesley, and *Numerical Recipes* by Press, Flannery, Teukolski, and Vetterling, Cambridge University Press.

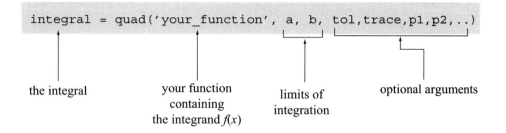

To use `quadl`, you just replace `quad` with `quadl` in the syntax. As shown in the syntax, both functions require you to supply the integrand as a user-written function. The optional input argument `tol` specifies absolute tolerance (the default value is 10^{-6}). A nonzero value of the other optional argument, `trace`, shows some intermediate calculations (see on-line help) at each step. The optional arguments `p1, p2`, etc., are simply passed on to the user-defined function as input arguments in addition to x.

The steps involved in numerical integration using these built-in functions are:

Step 1: Write a function that returns the value of the integrand $f(x)$ given the value of x. Your function should be able to accept the input x as a vector and, correspondingly, produce the value of the integrand (the output) as a vector.

Step 2: Decide which function to use—`quad` or `quadl` (`quad` is faster but less accurate than `quadl`). Find the integral. Decide if you need to change the default tolerance—`tol` $= 10^{-6}$. As a general guideline, if you use `quad` and you are not happy with the answer, use `quadl` before you start fiddling with the tolerance.

Example:

Let us compute the following integral

$$\int_{1/2}^{3/2} e^{-x^2} dx.$$

This integral is closely related to the error function, *erf*. In fact,

$$\int_0^x e^{-x^2} dx = \frac{\sqrt{\pi}}{2}\mathrm{erf}(x).$$

Because MATLAB also provides the error function, `erf`, as a built-in function, we can evaluate our integral in closed form (in terms of the error function) and compare the results from numerical integration. Let us follow the steps outlined previously.

Step 1: Here is the function that evaluates the integrand at given x (x is allowed to be a vector).

```
function y = erfcousin(x);
% ERFCOUSIN function to evaluate exp(-x^2).
y = exp(-x.^2);     % the array operator .^ is used for vector x
```

Step 2: Let us use quad with its simplest syntax:

```
>> y = quad('erfcousin',1/2,3/2)      % here a = 1/2, b = 3/2

y =

    0.3949
```

The exact result for the integral, up to 10 decimal places, is 0.3949073872. In the preceding example, we have used the default tolerance. Here is a table showing the results of a few experiments with this integral. We use both quad and quadl for integration. For quad, we tabulate results with different tolerances. In the table, we list the value of the integral, % error, and the number of function evaluations,[5] *F-evals* (a measure of computations involved). As you can see, quadl gives quite an accurate solution with just the default tolerance. Even quad does very well and gives only 0.0000014% error with the default tolerance and just about the same computational effort as quadl.

Function	tol	Answer	% Error	F-evals
quad	default	0.3949073927	1.3907×10^{-6}	17
	10^{-7}	0.3949073894	5.5506×10^{-7}	25
	10^{-8}	0.3949073873	2.4369×10^{-8}	33
quadl	default	0.3949073875	8.0616×10^{-8}	18

5.4.1 Double integration

To evaluate integrals of the form

$$\int_{y_{\min}}^{y_{\max}} \int_{x_{\min}}^{x_{\max}} f(x,y) \; dx \; dy$$

MATLAB provides a function dblquad. The calling syntax for dblquad is

I = dblquad('fxy_fun', xmin, xmax, ymin, ymax, *tol*, *@method*)

For on-line help type:
`help dblquad`

where *tol* and *method* are optional input arguments. The optional argument *tol* specifies tolerance (default is 10^{-6}), as previously discussed for 1-D integration, and *method* specifies a choice that the user makes about the method of integration to be used, e.g., quad or quadl. The default method is quad. The user-defined integrand function, fxy_fun, must be written such that it can accept a vector x and a scalar y while evaluating the integrand.

---

[5]The number of function evaluations can be obtained by including an optional output argument in the call syntax of quad: [y,fnevals]=quad('erfcousin', ...).

Because the double integrals are performed by carrying out 1-D integrals in one direction while holding the other variable constant and then repeating the procedure in the second direction, both `quad` and `quadl` are valid choices for the *method*. In addition, you could specify another *method* that you have programmed in a function, but your function must have the same input and output arguments, and in the same order, as `quad`.

Example:

Let us compute the following integral

$$I = \int_{-1}^{1} \int_{0}^{2} 1 - 6x^2 y \ dx \ dy.$$

It is fairly simple to verify analytically that $I = 4$. Let us see how `dblquad` performs on this integral.

```
>>  F = inline('1-6*x.^2*y');
>>  I = dblquad(F,0,2,-1,1)

I =

    4.0000
```

Create the integrand as an inline function. Note that x is taken as a vector argument. Next, run `dblquad` with default *tol* and *method*.

Note that we get the exact result just with default tolerance and the default lower-order method, `quad`. You can verify that the higher-order method `quadl` gives the same result by executing the command

```
I = dblquad(F,0,2,-1,1,[],@quadl)
```

Nonrectangular domains

The limits of integration used by `dblquad` are constants ($a, b, c,$ and d). Thus, the domain of integration is a rectangle. What if the inner limits of integration are variable rather than constant (i.e., the domain of integration is not a rectangle)? It is possible to do a change of variables such that the limits of integration become constant. See a book on numerical methods[6] for a description of the method. Once you convert the limits to constants, you can use `dblquad` to evaluate the integral. Alternatively, you can still use a rectangular domain enclosing the domain of interest and set the value of the integrand to be zero outside the domain of interest using relational operators. Let us take a simple example.

Example: Evaluate

$$\int_{0}^{1} \int_{y}^{1} x^2 e^{xy} \ dx \ dy.$$

The domain of integration here is a triangular region, say D, because x varies from $x = y$ to $x = 1$. We can still use `dblquad` over a rectangular region that contains

---

[6]For example, *Applied Numerical Analysis* by Gerald and Wheatley, Addison Wesley.

this triangular region by setting the integrand to be zero outside the triangular region. How do we do that? Well, we can multiply the integrand with another relational expression that returns 0 when $(x, y) \notin D$ and returns 1 if $(x, y) \in D$. This is simple enough; multiply the integrand with y-x<=0. This expression returns 0 when x is to the left of the line $y = x$ and returns 1 when x is to the right of $y = x$. Now we are ready to evaluate the integral.

```
F = inline('x.^2.*exp(x*y) .* (y-x<=0)');   % create the integrand
I = dblquad(F,0,1,0,1)             % integrate over the rectangular domain
```

The given integral happens to be easy enough so that we can integrate analytically and find that $I = e/2 - 1$. The following table shows how dblquad performs in this case.

Function	tol	Answer	% Error
default	default	0.35915942277049	0.0052
	10^{-8}	0.35914291876435	0.0005
quadl	default	0.35914850908091	0.0021
	10^{-8}	0.35914696325471	0.0017

Note that with default tolerance the higher-order method (quadl) performs better than the lower-order default method (quad), but at tighter tolerances there does not seem to be a clear winner.

5.5 Ordinary Differential Equations

For on-line help type:
`help funfun`

There is a separate suite of ordinary differential equation solvers in MATLAB. Long ago, MATLAB used to have just two built-in functions for solution of ODEs—ode23 and ode45. Now there are several additional functions that can also handle stiff equations. Although the various choices now available have increased the versatility of MATLAB's ODE solving capability, ode23 and ode45 remain the workhorses of the suite. In the new version of MATLAB, even ode23 and ode45 have changed and become more versatile. The versatility, however, comes with a cost; the more you want from these functions, the more you have to understand about their complex input structure. At this point, we will not go into such intricate details. We will look at the most straightforward and, perhaps, the most used form for the MATLAB's ODE solvers.

The functions ode23 and ode45 are implementations of second-/third-order and fourth-/fifth-order Runge-Kutta methods, respectively. Solving most ODEs using these functions in their simplest form (without any optional arguments) involves the following four steps:

1. **Write the differential equation(s) as a set of first-order ODEs.** For ODEs of order ≥ 2, this step involves introducing new variables and recasting the original equation(s) in terms of first-order ODEs in the new variables.

general Form

Basically, you need the equation in the vector form $\dot{\mathbf{x}} = \mathbf{f}(\mathbf{x},t)$, where $\mathbf{x} = [x_1 \quad x_2 \quad \ldots \quad x_n]^T$. In expanded form, the equation is:

$$\left\{\begin{array}{c} \dot{x}_1 \\ \dot{x}_2 \\ \vdots \\ \dot{x}_n \end{array}\right\} = \left\{\begin{array}{c} f_1(x_1, x_2, \ldots, x_n, t) \\ f_2(x_1, x_2, \ldots, x_n, t) \\ \vdots \\ f_n(x_1, x_2, \ldots, x_n, t) \end{array}\right\}.$$

2. **Write a function to compute the state derivative.** The state derivative $\dot{\mathbf{x}}$ is the vector of derivatives $\dot{x}_1, \dot{x}_2, \ldots, \dot{x}_n$. Therefore, you have to write a function that computes $f_1, f_2, \ldots, f_n$, given the input $(\mathbf{x}, t)$ where $\mathbf{x}$ is a column vector, that is, $\mathbf{x} = [x_1 \quad x_2 \quad \ldots \quad x_n]^T$. Your function must return the state derivative $\dot{\mathbf{x}}$ as a column vector.

3. **Use the built-in ODE solvers ode23 or ode45 to solve the equations.** Your function written in Step 2 is used as an input to ode23 or ode45. The syntax of use of ode23 is shown next. To use ode45 just replace ode23 with ode45.

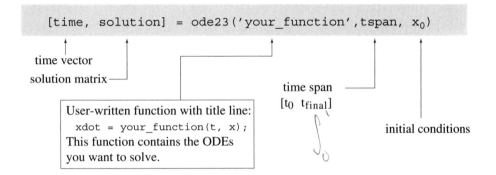

4. **Extract the desired variables from the output and interpret the results.** For a system of n equations, the output matrix solution contains n columns. You need to understand which column corresponds to which variable in order to extract the correct column, if you want to plot a variable with respect to, say, the independent variable time.

Here are two examples.

5.5.1 Example 1: A first-order linear ODE

Solve the first-order linear differential equation

$$\frac{dx}{dt} = x + t \tag{5.3}$$

with the initial condition

$$x(0) = 0.$$

Step 1: Write the equation(s) as a system of first-order equations: The given equation is already a first-order equation. No change is required.

$$\dot{x} = x + t.$$

Step 2: Write a function to compute the new derivatives: The function should return $\dot{x}$ given x and t. Here is the function:

```
function xdot = simpode(t,x);
% SIMPODE: computes xdot = x+t.
% call syntax:  xdot = simpode(t,x);
xdot = x + t;
```

Write and save it as an M-file named **simpode.m**.

Step 3: Use ode23 to compute the solution: The commands as typed in the command window are shown next. These commands could instead be part of a script file or even another MATLAB function. Note that we have not used the optional arguments `tol` or `trace`.

```
>>  tspan = [0  2];                 % specify time span
>>  x0 = 0;                         % specify x0
>> [t,x] = ode23('simpode',tspan,x0);  % now execute ode23
```

Step 4: Extract and interpret results: The output variables t and x contain results—t is a vector containing all discrete points of time at which the solution was obtained, and x contains the values of the variable x at those instances of time. Let us see the solution graphically:

```
>> plot(t,x)                        % plot t vs. x
>> xlabel('t')                      % label x-axis
>> ylabel('x')                      % label y-axis
```

The plot generated by the preceding commands is shown in Fig. 5.9.

5.5.2 Example 2: A second-order nonlinear ODE

Solve the equation of motion of a nonlinear pendulum

$$\ddot{\theta} + \omega^2 \sin\theta = 0 \quad \Rightarrow \quad \ddot{\theta} = -\omega^2 \sin\theta \tag{5.4}$$

with the initial conditions

$$\theta(0) = 1, \quad \dot{\theta}(0) = 0.$$

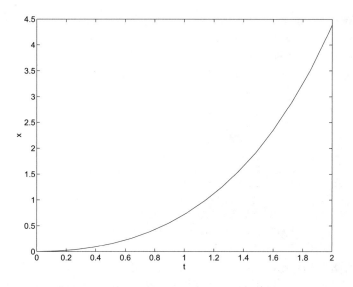

Figure 5.9: Numerical solution of the equation using ode23.

Step 1: Write the equation(s) as a system of first-order equations: The given equation is a second-order ODE. To recast it as a system of two first-order equations (an nth-order equation reduces to a set of n first-order equations), let us introduce two new variables.

Let $z_1 = \theta$ and $z_2 = \dot{\theta}$. Then $\dot{z}_1 = \dot{\theta} = z_2$ and $\dot{z}_2 = \ddot{\theta} = -\omega^2 \sin(z_1)$. Now eqn. (5.4) may be written in vector form as

$$\left\{ \begin{array}{c} \dot{z}_1 \\ \dot{z}_2 \end{array} \right\} = \left\{ \begin{array}{c} z_2 \\ -\omega^2 \sin(z_1) \end{array} \right\}.$$

We may write this equation in vector form as

$$\dot{\mathbf{z}} = \mathbf{f}(\mathbf{z})$$

where $\dot{\mathbf{z}}, \mathbf{z}$, and $\mathbf{f}(\mathbf{z})$ are vectors with two components each:

$$\mathbf{z} = \left\{ \begin{array}{c} z_1 \\ z_2 \end{array} \right\}, \qquad \dot{\mathbf{z}} = \left\{ \begin{array}{c} \dot{z}_1 \\ \dot{z}_2 \end{array} \right\} = \left\{ \begin{array}{c} z_2 \\ -\omega^2 \sin z_1 \end{array} \right\} = \mathbf{f}(\mathbf{z}).$$

This is a special case of $\dot{\mathbf{z}} = \mathbf{f}(t, \mathbf{z})$ where $\mathbf{f}$ does not depend on t. In general, $\mathbf{f}$ may contain time-dependent components as well. For example, we could have a driven pendulum, given by

$$\ddot{\theta} + \omega^2 \sin \theta = A \cos \Omega t.$$

In this case, we would write

$$\mathbf{f}(\mathbf{z}) \equiv \dot{\mathbf{z}} = \left\{ \begin{array}{c} z_2 \\ -\omega^2 \sin(z_1) + A \cos(\Omega t) \end{array} \right\}.$$

Step 2: Write a function to compute the new state derivative: We need to write a function that, given the scalar time t and vector $\mathbf{z}$ as input, returns the time derivative vector $\dot{\mathbf{z}}$ as output. In addition, the state derivative vector $\dot{\mathbf{z}}$ must be a column vector. Here is a function that serves the purpose.

```
function zdot = pend(t,z);
% Call syntax:  zdot = pend(t,z);
%  Inputs are:  t = time
%               z = [z(1); z(2)] = [theta; thetadot]
%  Output is:   zdot = [z(2); -w^2*sin z(1)]
wsq = 1.56;     % specify a value of w^2
zdot = [z(2); -wsq*sin(z(1))]
```

Note that z(1) and z(2) refer to the first and second elements of vector $\mathbf{z}$. Do not forget to save the function pend as an M-file to make it accessible to MATLAB.

Step 3: Use ode23 or ode45 for solution: Now, let us write a script file that solves the system and plots the results. Remember that the output $\mathbf{z}$ contains two columns: z_1, which is actually θ, and z_2, which is $\dot{\theta}$. Here is a script file that executes ode23, extracts the displacement and velocity in vectors x and y, and plots them against time as well as in the phase plane.

```
tspan = [0 20]; z0 = [1;0];     % assign values to tspan, z0
[t,z] = ode23('pend',tspan,z0); % run ode23
x = z(:,1); y = z(:,2);         % x=column 1 of z, y=column 2
plot(t,x,t,y)                   % plot t vs x and t vs y
xlabel('t'), ylabel('x and y')  % add axis labels
figure(2)                       % open a new figure window
plot(x,y)                       % plot phase portrait
xlabel('Displacement'), ylabel('Velocity')
title('Phase Plane Plot')       % put a title
```

Step 4: Extract and interpret results: The desired variables have already been extracted and plotted by the script file in Step 3. The plots obtained are shown in Fig. 5.10 and Fig. 5.11.

5.5.3 ode23 versus ode45

For solving most initial value problems, you use either ode23 or ode45. Which one should you choose? In general, ode23 is quicker but less accurate than ode45. However, the actual performance also depends on the problem. As the following table shows, ode45 is perhaps a better choice. We solved the two simple ODEs discussed in Sections 5.5.1 and 5.5.2 with these functions and noted the number of successful steps (*good steps*), number of failed steps (*bad steps*), and number of function evaluations (*f-evals*). Solutions were obtained with different *relative tolerances*. What does this tolerance mean? How do you set it? We discuss this in the following sections (Sections 5.5.4 and 5.5.5).

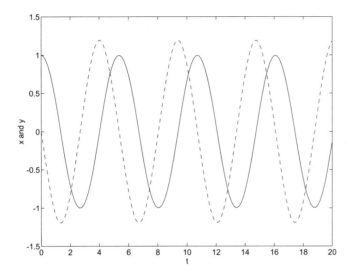

Figure 5.10: Displacement and velocity versus time plot of the pendulum.

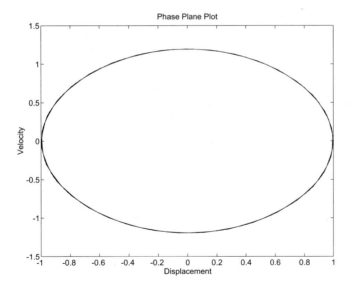

Figure 5.11: Displacement versus velocity plot of the pendulum.

RelTol	Solver	Good Steps	Bad Steps	f-evals
Solve $\dot{x} = x + t, \quad x(0) = 0, \quad 0 \le t \le 2$				
default	ode23	17	3	61
default	ode45	10	0	61
Solve $\ddot{\theta} = -\omega^2 \sin\theta, \quad \theta(0) = 1, \dot{\theta}(0) = 0, \quad 0 \le t \le 20$				
default	ode23	114	13	382
default	ode45	33	0	199
10^{-4}	ode23	251	15	799
10^{-4}	ode45	49	1	301
10^{-5}	ode23	528	0	1585
10^{-5}	ode45	80	10	541
10^{-6}	ode23	805	0	2416
10^{-6}	ode45	104	0	625

5.5.4   Specifying tolerance

Many functions in the `funfun` category provide the user with a choice of specifying a *tolerance* as an optional input argument. A tolerance is a small positive number that governs the error of some appropriate kind in the computations of that function. For example, the ODE solvers `ode23` or `ode45` use the tolerance in computing the step size based on the error estimate of the computed solution at the current step. So what value of the tolerance should you specify? To make your decision even harder, there are two of them—relative tolerance and absolute tolerance.

The relative tolerance, in general, controls the number of correct digits in the solution of a component. Thus, if you want the solution to have k digit accuracy, you should set the relative tolerance to 10^{-k}. This is because the error estimated in the solution of a component y_i is compared with the value *Relative Tolerance* $\times |y_i|$. In fact, MATLAB uses a combination of both the relative tolerance and the absolute tolerance. For example, in ODE solvers, the estimated local error e_i in the solution component y_i at any step is required to satisfy the relationship

$$e_i \le \max(RelTol \times |y_i|, AbsTol).$$

The absolute tolerance specifies the threshold level for any component of the solution below which the computed values of the component have no guarantee of any accuracy. From the error relationship shown previously, it should be clear that specifying a very low value of relative tolerance, below the absolute tolerance, may not lead to more accurate solutions.

The absolute tolerance can also be specified as a vector, in which case it specifies the threshold values of each solution component separately. The value of a solution component below the corresponding absolute tolerance is almost meaningless.

5.5.5   The ODE suite

The ODE suite of MATLAB consists of several other solvers (mostly for stiff ODEs and differential algebraic equations (DAEs) of index 1) and utility functions. Even the old faithfuls, `ode23` and `ode45`, have been rewritten for better performance. The following list shows some of the functions and utilities available in MATLAB. You should see the on-line help on these functions before using them.

`ode45`	nonstiff solver based on fourth-/fifth-order Runge-Kutta method,
`ode15s`	stiff solver and DAE solver based on a variable order method,
`ode23`	nonstiff solver based on second-/third-order Runge-Kutta method,
`ode113`	nonstiff solver based on variable order, Adams-Bashforth-Moulton methods,
`ode23t`	stiff solver for moderately stiff equations and DAEs, based on trapezoidal rule,
`ode23s`	stiff solver based on a low order, numerical differentiation formula (NDF), and
`ode23tb`	stiff solver based on a low-order method.

Utility functions

`odefile`	a help file to guide you with the syntax of your ODE function,
`odeset`	a function that sets various **options** for the solvers,
`odeget`	a function that gets various **options** parameters,
`odeplot`	a function that is specified in **options** for time series plots, and
`odephas2`	a function that is specified in **options** for 2-D phase plots.

Because there are so many choices for solvers, there is likely to be some confusion about which one to choose. If you do not have stiff equations, your choices are limited to `ode23`, `ode45`, and `ode113`. A general rule of thumb is that `ode45` will give you satisfactory results for most problems, so try it first. If you suspect the equations to be stiff, try `ode15s` first from the list of stiff solvers. If that doesn't work, learn about stiffness, understand your equations better, and find out more about the stiff solvers in the ODE suite before you start trying them out.

One nice feature of the entire ODE suite is that all solvers use the same syntax. We discussed the simplest form of the syntax in the previous section. Now we will look at the full syntax of these functions:

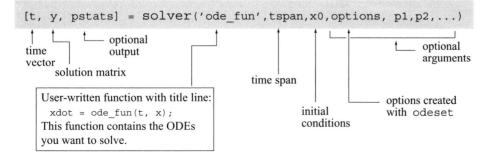

There are three new optional arguments here—one in the output, stats, and two in the input, options and p1, p2, $\cdots$, pn. Here is a brief description of these arguments.

pstats: It is a six-element-long vector that lists the performance statistics of the solvers. The first three elements are (i) number of successful steps, (ii) number of failed steps, and (iii) number of function evaluations. The next three elements are related to performance indices of stiff solvers.

options: It is a structure that is created with the function odeset using the command

options = odeset('*name1*',value1,'*name2*',value2,$\cdots$)

where *name* refers to the name of the optional argument that you want to set and value refers to the corresponding value of the argument. The list of the arguments that you can set with odeset includes relative tolerance (reltol), absolute tolerance (abstol), choice of output function (outputfcn), performance statistics (stats). There are many more optional arguments. See the on-line help on odeset. The most commonly used arguments, perhaps, are reltol, abstol, stats, and outputfcn. When stats is 'on', the solver performance statistics are displayed at the end. It is an alternative[7] to specifying the optional output argument pstats discussed earlier. The argument outputfcn can be assigned any of the output functions, 'odeplot', 'odephas2', 'odephas3', or 'odeprint', to generate a time series plot, phase plot in 2-D or 3-D, or show the computed solution on the screen, as the solution proceeds.

As an example, suppose we want to set the relative tolerance to 10^{-6}, absolute tolerance to 10^{-8}, see the performance statistics, and have MATLAB plot the time series of the solution at each time step as the solution is computed. We set the options structure as follows:

options = odeset('reltol',1e-6,'abstol',1e-8,...
                    'stats','on', 'outputfcn','odeplot');

p1, p2, ..., pn: These are optional arguments that are directly passed on to the user-written ODE function. In previous versions of MATLAB, the only arguments that the solvers could pass on to the ODE function were (t, x). Thus, if the user-written function needed other variables to be passed, they had to be either hardcoded or passed on through global declaration. To find out how to use these variables in the ODE function and how to write the list of input variables to incorporate these optional variables, please see the on-line help on odefile.

---

[7]Of course, if you specify the optional output argument, the statistics are saved in that variable, whereas with 'stats' 'on' in the options, you see the statistics but cannot access them later.

5.5.6 Event location

In solving initial value problems, usually the termination condition is specified in terms of the independent variable. As shown in the previous two examples, the solution stops when the independent variable t reaches a prescribed value t_{final} (specified as the second element of `tspan`). Thus, we obtain the solution for a certain time span. Sometimes, however, we need to stop the solution at a specified value of the dependent variable and we do not know when (at what value of t) the solution will reach there. For example, say we want to solve a projectile problem. We write the equations of motion (two ODEs for $\ddot{x}$ and $\ddot{y}$), and we would like to solve the equations to find *when* the projectile hits the target. If we do not know t_{final} apriori, what value of t_{final} should we specify in `tspan`? One thing we know is that the solution should stop when the target (some specified location x_{target} or y_{target}) is hit, that is, when x or y reach a particular value. When the dependent variables (here $x(t)$ or $y(t)$) reach some specified value, we call that an *event*, and the problem of finding the time of the event is called *event location*. Event location problems are solved by following the solution until the solution *crosses* the event, then backtracking to the time just before the event, and then taking smaller and suitably computed time steps to land exactly at the event. There are several strategies and algorithms for event location. In the literature, the event location problem is also referred to as *integrating across discontinuities*.

Fortunately, the ODE solvers in MATLAB have built-in ability to solve event location problems. Let us take a simple example—a projectile is thrown with initial speed v_o at an angle θ. We want to find out (i) the instant when the projectile hits the ground, (ii) the range, and (iii) the trajectory of the projectile. The ODEs governing the motion of the projectile are

$$\ddot{x} = 0$$
$$\ddot{y} = -g$$

The initial conditions are $x(0) = y(0) = 0$, $\dot{x}(0) = v_0 \cos\theta$, and $\dot{y}(0) = v_0 \sin\theta$. The equations of motion are to be integrated until $y = 0$ (the projectile hits the ground).

Before we write a function to code our equations, we need to convert them into a set of first-order equations:

Let $x = x_1, \dot{x} = x_2, y = x_3$, and $\dot{y} = x_4$, then $\dot{x}_1 = x_2, \dot{x}_2(= \ddot{x}) = 0, \dot{x}_3 = x_4$, and $\dot{x}_4(= \ddot{y}) = -g$. In vector form:

$$\begin{Bmatrix} \dot{x}_1 \\ \dot{x}_2 \\ \dot{x}_3 \\ \dot{x}_4 \end{Bmatrix} = \begin{Bmatrix} x_2 \\ 0 \\ x_4 \\ -g \end{Bmatrix} \tag{5.5}$$

Now, eqn. (5.5) is ready for coding in a function.

Event specification

In this example, the event takes place at $y = 0$. But if we somehow instruct the solver to stop when $y = 0$, our projectile will never take off because the solver will stop right at the start, since $y(0) = 0$. Therefore, we have to make the event detection more robust by adding some more condition(s). In this example, we can use the fact that the vertical component of the velocity will be negative when the projectile hits the ground, i.e., $\dot{y} < 0$. Thus, the conditions for the event detection are

$$y = 0, \qquad \dot{y} < 0.$$

Once the event is detected, we have to make another decision—what to do next. We could stop the solution (as we desire in the present example) or we could simply note the event and continue with or without changing some variables (for example, a ball bouncing off a wall would require the detection of collision with the wall, application of collision laws to determine the reversed initial velocity, and then continuation of the solution). If the solution has to stop at the event, we will call the event *terminal*, or else *nonterminal*. Our event is terminal. This completes the specification of the event.

Solution with event detection

We will now show how to solve the projectile problem using MATLAB's solver ode45 such that the solver automatically detects the collision of the projectile with the ground and halts the program. The steps involved are as follows.

Step 1: Tell the solver to watch out for *events*: We need to tell the solver that we want it to look for *events* that we are going to specify. This is done by setting the 'events' flag 'on' in the options with the function odeset:

$$\texttt{options = odeset('events','on');}$$

Of course, when we run the solver (see Step 3), we will use options as an optional input argument.

Step 2: Write your function to give *event* information: We need to write the ODE function with some additional features now. The MATLAB solver is going to call this function with an extra input variable, flag, that tells the ODE function what kind of output is desired. When the solver calls the ODE function with the flag set to events, it looks for three output quantities from the ODE function:

1. value: a vector of values of those variables for which a zero crossing defines the *event*. In our example, the event (collision with the ground) is defined by $y = 0$, i.e., a zero crossing of the y-value. Thus, in our case, value=x(3) (the value of y).

2. `isterminal`: a vector of ones and zeros specifying whether occurrence of an event is terminal (stops the solution) or not. For each element in the vector `value`, a 0 or a 1 is required in the vector `isterminal`. In our example, we have only one element in `value` (`x(3)` or y), therefore, we need just one element in `isterminal`. Because we would like to stop the program at $y = 0$, the zero crossing of y is terminal. Therefore, we set `isterminal=1`.

3. `direction`: a vector of the same length as `value`, specifying the direction of zero crossing for each element in `value`: a -1 for negative value, a 1 for positive value, and a 0 for *I-don't-care* value. In our example, `direction=-1` will imply that the zero crossing of y is a valid event only when the value of y is decreasing (y crosses zero with $\dot{y} < 0$). As you can see, specification of `direction` prevents the solver from aborting the solution at the outset when $y = 0$ but $\dot{y} \not< 0$.

Of course, your function should provide the event information, in terms of these three vectors as output, but only when asked. Thus, your function should check the value of the `flag` and output the three vectors if the `flag` is set to `events`, otherwise it should output the usual derivative vector `xdot`. We can implement this conditional output using `switch` (you can do it using `if-elseif` construction too), as shown in the following example function.

```
function [value,isterminal,dircn] = proj(t,z,flag);
% PROJ: ODE for projectile motion with event detection
g = 9.81;                    % specify constant g
if nargin<3 | isempty(flag) % if no flag or empty flag
   value = [z(2); 0; z(4); -g];
else
   switch flag               % see Section 4.3.4 for 'switch'
     case 'events'
       value = z(3);         % 'value' is zero crossing for y
       isterminal = 1;       % 'isterminal': y=0 is terminal
       dircn = -1;           % 'direction': ydot < 0
     otherwise
       error('function not programmed for this event');
   end                       % end of switch
end                          % end of if
```

Step 3: Run the solver with appropriate input and output: Clearly, we need to include `options` in the input list. In the output list, we have the choice of including some extra variables:

te   times at which the specified events occur,

ze   solutions (values of state variables) at *te*, and

ie   indices of events that occurred at *te*.

So, now let us solve the projectile problem. Our ODE function `proj` is ready. Here is a script file that specifies the required input variables and executes `ode45` with appropriate output variables. The trajectory is shown in Fig. 5.12, along with some information about range and time of flight.

```
% RUNPROJ: a script file to run the projectile example
tspan=[0 2];                    % specify time span, [t0 tfinal]
v0=5; theta = pi/4;             % theta must be in radians
z=[0; v0*cos(theta); 0; v0*sin(theta)];
                                % specify initial conditions
options = odeset('events', 'on');
                                % set the events switch on
[t,z,te,ze,ie] = ode45('proj',tspan,z0,options);
                                % run ode45 with options
x = z(:,1);   y = z(:,3);       % separate out x and y
plot(x,y), axis('equal')        % plot trajectory
xlabel('x'), ylabel('y'),       % label axes
title('Projectile Trajectory')  % write a title
info = ['Range (x-value) = ', num2str(ze(1)), ' m';
        'Time of flight  = ', num2str(te), ' s'];
                                % create strings of information
text([1;1],[0;-.1], info)       % print Range and Time on plot
```

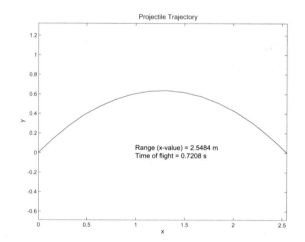

Figure 5.12: Trajectory of a projectile obtained by integrating equations of motion with `ode45` and using the event detection facility of the solver.

5.6 Nonlinear Algebraic Equations

The MATLAB function `fzero` solves nonlinear equations involving one variable (see also function `solve` in Chapter 8 if you have the Symbolic Math Toolbox). You can use `fzero` by proceeding with the following three steps.

1. **Write the equation in the standard form:**

$$f(x) = 0.$$

This step usually involves trivial rearrangement of the given equation. In this form, solving the equation and *finding a zero of* $f(x)$ are equivalent.

2. **Write a function that computes** $f(x)$: The function should return the value of $f(x)$ at any given x.

3. **Use the built-in function** `fzero` **to find the solution:** `fzero` requires an initial guess and returns the value of x closest to the guess at which $f(x)$ is zero. The function written in Step 2 is used as an input to the function `fzero`. The call syntax of `fzero` is:

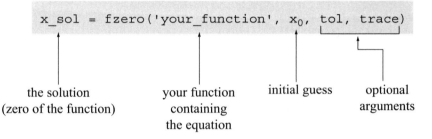

the solution   your function   initial guess   optional
(zero of the function)   containing   arguments
the equation

Example: A transcendental equation

Solve the following transcendental equation:

$$\sin x = e^x - 5.$$

Step 1: Write the equation in standard form: Rearrange the equation as:
$\sin(x) - e^x + 5 = 0 \quad \Rightarrow \quad f(x) = \sin(x) - e^x + 5.$

Step 2: Write a function that computes $f(x)$**:** This is easy enough:

```
function f = transf(x);
% TRANSF: computes f(x) = sin(x)-exp(x)+5.
% call syntax:  f = transf(x);
f = sin(x) - exp(x) + 5;
```

Write and save the function as an M-file named **transf.m**.

Step 3: Use `fzero` **to find the solution:** The commands as typed in the command window are shown next. The result obtained is also shown. Note that we have not used the *optional arguments* `tol` or `trace`.

```
>> x = fzero('transf',1)          % initial guess x0=1.

x =

    1.7878
```

To check the result, we can plug the value back into the equation or plot $f(x)$ and see if the answer looks right. See the plot of the function in Fig. 5.13.

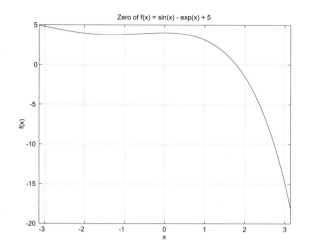

Figure 5.13: `fzero` locates the zero of this function at $x = 1.7878$.

5.6.1 Roots of polynomials

You can also find the zeros of a polynomial equation (e.g., $x^5 - 3x^3 + x^2 - 9 = 0$) with `fzero`. However, `fzero` locates the root closest to the initial guess; it does not give all roots. To find all roots of a polynomial equation, use the built-in function `roots`. This function requires the coefficients of the powers of x (in decreasing order), including the constant (the coefficient of x^0), in a vector and gives all the roots as the output. Thus, for the polynomial equation

$$x^5 - 3x^3 + x^2 - 9 = 0,$$

the coefficients are

$$\underbrace{(1)}_{C_1} x^5 + \underbrace{(0)}_{C_2} x^4 + \underbrace{(-3)}_{C_3} x^3 + \underbrace{(1)}_{C_4} x^2 + \underbrace{(0)}_{C_5} x + \underbrace{(-9)}_{C_6} = 0$$

or $C = [1 \quad 0 \quad -3 \quad 1 \quad 0 \quad -9]$. Thus, the coefficient vector is always of length $n + 1$ where n is the order of the polynomial. In fact, it is from the length of the vector that MATLAB figures out the polynomial order. Now here is how you find the roots:

```
>> c = [1 0 -3 1 0 -9];              % Coefficient vector C
>> roots(c)

ans =

   1.9316
   0.5898 + 1.1934i
   0.5898 - 1.1934i
  -1.5556 + 0.4574i
  -1.5556 - 0.4574i
```

Finding multiple roots of a nonpolynomial or finding roots of functions of several variables is a more advanced problem.

We need to mention some new functions that have been added in MATLAB (from version 6 onward) for facilitating optimization calculations. In particular, `fminbnd` and `fminsearch` have been added to find the unconstrained minimum of a nonlinear function. See on-line help for more details.

5.7 Advanced Topics

The topics covered in the preceding sections of this chapter are far from exhaustive in what you can do readily with MATLAB. These topics have been selected carefully to introduce you to various applications that you are likely to use frequently in your work. Once you gain a little bit of experience and some confidence, you can explore most of the advanced features of the functions introduced as well as several functions for more complex applications on your own, with the help of on-line documentation.

MATLAB provides some new functions for solving two-point boundary value problems, simple partial differential equations, and nonlinear function minimization problems. In particular, we mention the following functions.

`dde23` added in MATLAB 7, this function solves delay differential equations (DAEs) with constant delays; that is, differential equations of the form

$$\dot{x} = f\big(t, x(t), x(t - \tau_1), x(t - \tau_2), ..., x(t - \tau_k)\big)$$

where τ_j's represent constant delays.

`ode15i` solves implicit ODEs and DAEs of index 1 with the helper function `odeic` for evaluating consistent initial conditions.

`bvp4c` solves two-point boundary value problems (BVP) defined by a set of ODEs of the form $y' = f(x, y)$, and its boundary conditions $y(a)$ and $y(b)$ over the interval $[a, b]$. The user has to write two functions—one that specifies the equations and the other that specifies the boundary conditions. Users can set several options for initiating the solutions. To see an example, try executing the built-in example `twobvp` and learn how to program your BVP by following this example (to see the functions required for `twobvp`, type `type twoode.m` and `type twobc.m`).

`pdepe` solves simple parabolic and elliptic partial differential equations (PDEs) of a single dependent variable. This is a rather restricted utility function. However, for those who need to solve PDEs frequently, there is the Partial Differential Equation Toolbox.

EXERCISES

1. **Linear algebraic equations:** Find the solution of the following set of linear algebraic equations, as advised below.

$$x + 2y + 3z = 1$$
$$3x + 3y + 4z = 1$$
$$2x + 3y + 3z = 2.$$

- Write the equation in matrix form and solve for $\mathbf{x} = [x\ y\ z]^T$ using the left division \.
- Find the solution again using the function `rref` on the augmented matrix.
- Can you use the LU decomposition to find the solution? [Hint: Because $[\mathbf{LU}]\mathbf{x} = \mathbf{b}$, let $[\mathbf{U}]\mathbf{x} = \mathbf{y}$, so that $[\mathbf{L}]\mathbf{y} = \mathbf{b}$. Now, first solve for $\mathbf{y}$ and then for $\mathbf{x}$.]

2. **Eigenvalues and eigenvectors:** Consider the following matrix.

$$\mathbf{A} = \begin{bmatrix} 3 & -3 & 4 \\ 2 & -3 & 4 \\ 0 & -1 & 1 \end{bmatrix}$$

- Find the eigenvalues and eigenvectors of $\mathbf{A}$.
- Show, by computation, that the eigenvalues of $\mathbf{A}^2$ are square of the eigenvalues of $\mathbf{A}$.
- Compute the square of the eigenvalues of $\mathbf{A}^2$. You have now obtained the eigenvalues of $\mathbf{A}^4$. From these eigenvalues, can you guess the structure of $\mathbf{A}^4$?
- Compute $\mathbf{A}^4$. Can you compute $\mathbf{A}^{-1}$ without using the `inv` function?

3. **Linear and quadratic curve fits:** The following data is given to you.

$$\begin{bmatrix} x \\ y \end{bmatrix} = \begin{bmatrix} 0 & 0.15 & 0.2 & 0.30 & 0.4 & 0.52 & 0.6 & 0.70 & 0.8 & 0.90 & 1 \\ 0 & 1.61 & 2.2 & 3.45 & 4.8 & 6.55 & 7.8 & 9.45 & 11.2 & 13.05 & 15 \end{bmatrix}$$

- Enter vectors x and y and plot the raw data using `plot(x,y,'o')`.
- Click on the figure window. Select **Basic Fitting** from the **Tools** menu of the figure window. Once you get the **Basic Fitting** window, check the box for **linear** fit. In addition, check the boxes for **Show equations**, **Plot residuals** (select a *scatter* plot rather than a *bar* plot option), and **Show norm of residuals**.
- Now, do a quadratic fit by checking the **quadratic** fit box and all other boxes as you did for the linear case.
- Compare the two fits. Which fit is better? There isn't really a competition, is there?

4. **A nonlinear curve fit:** Enter the following experimental data in MATLAB workspace.

```
t = [0 1.40 2.79 4.19 5.58 6.98 8.38 9.77 11.17 12.57];
x = [0 1.49 0.399 -0.75 -0.42 0.32 0.32 -0.10 -0.21 0];
```

You are told that this data comes from measuring the displacement x of a damped oscillator at time instants t. The response x is, therefore, expected to have the following form:

$$x = Ce^{-\lambda_1 t}\sin(\lambda_2 t)$$

where the constants C, λ_1, and λ_2 are to be determined from the experimental data. Thus, your job is to fit a curve of the given form to the data and find the constants that give the best fit.

- First, see a built-in demo for nonlinear curve fits by typing `fitdemo`.
- Write a function that computes x given some initial guess of the values of the constants C, λ_1, and λ_2 as follows.

```
function er_norm = expsin(constants,t_data,x_data);
% given lambda, compute norm of the error in the fitted curve
t = t_data'; x = x_data' ;   % make sure t & x are columns
C = constants(1); l1 = constants(2); l2 = constants(3);
xnew = C * exp(-l1 * t) .* sin(l2 *t);   % evaluate your x
er_norm = norm(xnew - x) % compare with data and compute error
```

- Now you can find the best **constants** iteratively by minimizing the norm of the error with the function `fminsearch` as follows.

```
constants = [1 0.1 0.8];   % initial guess of [C, l1, l2]
constants = fminsearch('expsin',constants,[],t,x);
```

- Now that you have the *best fit* constants, evaluate your formula at the given t and plot the fitted curve along with the given data. That's it.

5. **Data statistics:** MATLAB provides a graphical tool for finding simple statistical measures on data. Consider the data given in Problem 3.

   - Plot the raw data with the marker `'o'`.
   - Select **Data Statistics** from the **Tools** pull-down menu of the figure window. A new window appears that shows the basic statistical measures of the data set.
   - To include any of these measures in your plot, simply check the appropriate box.

6. **Length of a curve through quadrature:** The length of a parametric curve defined by $x(t)$ and $y(t)$ over $a \leq t \leq b$ is given by the integral $\int_a^b \sqrt{(x')^2 + (y')^2}\, dt$ where $x' = dx/dt$ and $y' = dy/dt$. Find the length of a hypocycloid defined by $x(\theta) = a \cos^3 \theta$ and $y(\theta) = a \sin^3 \theta$ for $0 \leq \theta \leq \pi/2$. Take $a = 1$.

7. **A second-order linear ODE, the phenomenon of beats and resonance:** Consider the following simple linear ODE of second order.

$$\ddot{x} + x = F_0 \cos \omega t.$$

   - Convert the given equation to a set of first-order ODEs and program the set of equations in a function to be used for numerical solution with `ode45`.
   - Set $F_0 = 0$ and solve the system of equations with the initial conditions $x(0) = 0$, $\dot{x}(0) = 1$. By plotting the solution, make sure that you get a periodic solution. Can you find the period of this solution?
   - Let $F_0 = 1$ and $\omega = 0.9$. Starting with zero initial conditions ($x(0) = \dot{x}(0) = 0$), find the solutions x and $\dot{x}$ for $0 \leq t \leq 70$. Plot x against t. How is the amplitude of the rapidly oscillating solution modulated? You should see the beats phenomenon.
   - Let $F_0 = 1$ and $\omega = 1$. Again starting with zero initial conditions, find the solution for $0 \leq t \leq 40$. How does the amplitude of the solution, x, change with time in this case? This is the phenomenon of resonance.
   - Experiment with the solution of the system by adding a damping term, say $\zeta \dot{x}$, on the left-hand side of the equation.

6. Graphics

MATLAB includes good tools for visualization. Basic 2-D plots, fancy 3-D graphics with lighting and *colormaps*, complete user control of the graphics objects through *Handle Graphics* tools for designing sophisticated graphics user interfaces, and animation are now part of MATLAB. What is special about MATLAB's graphics facility is its ease of use and expandability. Commands for most garden-variety plotting are simple, easy to use, and intuitive. If you are not satisfied with what you get, you can control and manipulate virtually everything in the graphics window. This, however, requires an understanding of Handle Graphics, a system of low-level functions to manipulate graphics objects. In this section, we take you through the main features of MATLAB's graphics facilities.

6.1 Basic 2-D Plots

For on-line help type:
`help graph2d`

The most basic and perhaps most useful command for producing a 2-D plot is

$$\boxed{\texttt{plot}(\textit{xvalues, yvalues,'style-option'})}$$

where *xvalues* and *yvalues* are vectors containing the x- and y-coordinates of points on the graph and the *style-option* is an optional argument that specifies the color, the line style (e.g., solid, dashed, dotted), and the point-marker style (e.g., o, +, *). All three style options can be specified together. The two vectors *xvalues* and *yvalues* MUST have the same length. Unequal length of the two vectors is the most common source of error in the plot command. The `plot` function also works with a single-vector argument, in which case the elements of the vector are plotted against row or column indices. Thus, for two column vectors x and y each of length n,

`plot(x,y)`	plots y versus x with a solid line (the default line style),
`plot(x,y,'--')`	plots y versus x with a dashed line (more on this below), and
`plot(x)`	plots the elements of x against their row index.

6.1.1 Style options

The *style-option* in the plot command is a character string that consists of one, two, or three characters that specify the color and/or line style. There are several color, line, and marker style-options:

Color Style-option		Line Style-option		Marker Style-option	
y	yellow	–	solid	+	plus sign
m	magenta	--	dashed	o	circle
c	cyan	:	dotted	*	asterisk
r	red	-.	dash-dot	x	x-mark
g	green	none	no line	.	point
b	blue			ˆ	up triangle
w	white			s	square
k	black			d	diamond, etc.

The style-option is made up of the color option, the line option, the marker option, or a combination of them.

Examples:

`plot(x,y,'r')`	plots y versus x with a red solid line,
`plot(x,y,':')`	plots y versus x with a dotted line,
`plot(x,y,'b--')`	plots y versus x with a blue dashed line, and
`plot(x,y,'+')`	plots y versus x as unconnected points marked by +.

When no style-option is specified, MATLAB uses a blue solid line by default.

6.1.2 Labels, title, legend, and other text objects

Plots may be annotated with `xlabel`, `ylabel`, `title`, and `text` commands.

The first three commands take string arguments, whereas the last one requires three arguments—`text`(*x-coordinate, y-coordinate, 'text'*), where the coordinate values are taken from the current plot. Thus,

`xlabel('Pipe Length')`	labels the x-axis with `Pipe Length`,
`ylabel('Fluid Pressure')`	labels the y-axis with `Fluid Pressure`,
`title('Pressure Variation')`	titles the plot with `Pressure Variation`, and
`text(2,6,'Note this dip')`	writes "`Note this dip`" at the location (2.0,6.0) in the plot coordinates.

We have already seen an example of `xlabel`, `ylabel`, and `title` in Fig. 3.9. An example of `text` appears in Fig. 6.2. The arguments of `text`(*x,y, 'text'*) command may be vectors, in which case x and y must have the same length and *text* may be just one string or a vector of strings. If *text* is a vector, then it must have the same length as x. A useful variant of the `text` command is `gtext`, which only takes a string argument (a single string or a vector of strings) and lets the user specify the location of the text by clicking the mouse at the desired location in the graphics window.

For on-line help type:
`help graph2d`

Legend

Legends on plots can be produced using the Insert → legend button in the figure window toolbar (see Fig. 6.1) or with the `legend` command. The `legend` command produces a boxed legend on a plot, as shown, for example, in Fig. 6.3 on page 182. The `legend` command is quite versatile. It can take several optional arguments. The most commonly used forms of the command are listed here.

`legend(string1, string2, ..)`   produces legend using the text in
                        string1, string2, etc., as labels,
`legend(LineStyle1, string1, ..)` specifies the line style of each label,
`legend(.., pos)`   writes the legend outside the plot-frame
                        if *pos* = −1 and inside if *pos* = 0,
                        (there are other options for *pos* too), and
`legend off`   deletes the legend from the plot.

When MATLAB is asked to produce a legend, it tries to find a place on the plot where it can write the specified legend without running into lines, grids, and other graphics objects. The optional argument *pos* specifies the location of the legend box. `pos=1` places the legend in the upper right-hand corner (default), 2 in the upper left-hand corner, 3 in the lower left-hand corner, and 4 in the lower right-hand corner. The user, however, can move the legend at will with the mouse (click and drag). For more information, see the on-line help on `legend`.

6.1.3   Axis control, zoom in, and zoom out

Once a plot is generated, you can change the axes limits with the `axis` command. Typing

$$\boxed{\texttt{axis([xmin\ xmax\ ymin\ ymax])}}$$

changes the current axes limits to the specified new values *xmin* and *xmax* for the *x*-axis and *ymin* and *ymax* for the *y*-axis.

Examples:

`axis([-5 10 2 22]);`   sets the *x*-axis from −5 to 10, *y*-axis from 2 to 22,
`axy = [-5 10 2 22]; axis(axy);`   same as above, and
`ax = [-5 10]; ay=[2 22]; axis([ax ay]);`   same as above.

The `axis` command may thus be used to zoom in on a particular section of the plot or to zoom out.[1] There are also some useful predefined string arguments for the `axis` command:

`axis('equal')`   sets equal scale on both axes,
`axis('square')`   sets the default rectangular frame to a square,
`axis('normal')`   resets the axis to default values,
`axis('axis')`   freezes the current axes limits, and
`axis('off')`   removes the surrounding frame and the tick marks.

The `axis` command must come after the `plot` command to have the desired effect.

---

[1]There is also a `zoom` command that can be used to zoom in and zoom out using the mouse in the figure window. See the on-line help on `zoom`.

Semi-control of axes

It is possible to control only part of the axes limits and let MATLAB set the other limits automatically. This is achieved by specifying the desired limits in the `axis` command along with `inf` as the values of the limits that you would like to be set automatically. For example,

`axis([-5 10 -inf inf])`   sets the x-axis limits at -5 and 10 and lets the y-axis limits be set automatically, and

`axis([-5 inf -inf 22])`   sets the lower limit of the x-axis and the upper limit of the y-axis, and leaves the other two limits to be set automatically.

For on-line help type:
`help plotedit`
`help propedit`

6.1.4   Modifying plots with the plot editor

MATLAB provides an enhanced (over previous versions) interactive tool for modifying an existing plot. To activate this tool, go to the figure window and click on the left-leaning *arrow* in the menu bar (see Fig. 6.1). Now you can select and double-click (or right-click) on any object in the current plot to edit it. Double-clicking on the selected object brings up a property editor window where you can select and modify the current properties of the object. Other tools in the menu bar, e.g., text (marked by **A**), arrow, and line, lets you modify and annotate figures just like simple graphics packages do.

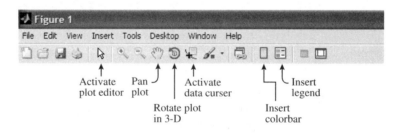

Figure 6.1: MATLAB provides interactive plot editing tools in the **Figure** window menu bar. Click on the white arrow to activate the plot editor. With the arrow selected, you can double-click on any graphics object in the figure window to open its properties box and edit the desired properties.

You can also activate the plot editor in the figure window by typing `plotedit` at the command prompt. You can activate the property editor by typing `propedit` at the command prompt. However, to make good use of the property editor, you must have some understanding of Handle Graphics. See Section 6.4 on page 205 for details.

6.1.5 Overlay plots

There are three different ways of generating overlay plots in MATLAB: the `plot`, `hold`, and `line` commands.

Method 1: Using the `plot` command to generate overlay plots

If the entire set of data is available, `plot` command with multiple arguments may be used to generate an overlay plot. For example, if we have three sets of data—$(x1, y1)$, $(x2, y2)$, and $(x3, y3)$—the command `plot(x1,y1, x2,y2,':', x3,y3,'o')` plots $(x1, y1)$ with a solid line, $(x2, y2)$ with a dotted line, and $(x3, y3)$ as unconnected points marked by small circles (o), all on the same graph (see Fig. 6.2 for example). Note that the vectors (xi, yi) must have the same length pairwise. If the length of all vectors is the same, then it is convenient to make a matrix of x vectors and a matrix of y vectors and then use the two matrices as the argument of the `plot` command. For example, if $x1$, $y1$, $x2$, $y2$, $x3$, and $y3$ are all column vectors of length n, then typing `X=[x1 x2 x3]; Y=[y1 y2 y3]; plot(X,Y)` produces a plot with three lines drawn in different colors. When `plot` command is used with matrix arguments, each column of the second argument matrix is plotted against the corresponding column of the first argument matrix.

Method 2: Using the `hold` command to generate overlay plots

Another way of making overlay plots is with the `hold` command. Invoking `hold` on at any point during a session freezes the current plot in the graphics window. All subsequent plots generated by the `plot` command are simply added to the existing plot. The following script file shows how to generate the same plot as in Fig. 6.2 by using the `hold` command.

```
% - Script file to generate an overlay plot with the hold  command -
x = linspace(0,2*pi,100);       % Generate vector x
y1 = sin(x);                    % Calculate y1
plot(x,y1)                      % Plot (x,y1) with solid line
hold on                         % Invoke hold for overlay plots
y2 = x; plot(x,y2,'--')         % Plot (x,y2) with dashed line
y3 = x - (x.^3)/6 + (x.^5)/120; % Calculate y3
plot(x,y3,'o')                  % Plot (x,y3) as pts. marked by 'o'
axis([0 5 -1 5])                % Zoom in with new axis limits
hold off                        % Clear hold command
```

The `hold` command is useful for overlay plots when the entire data set to be plotted is not available at the same time. You should use this command if you want to keep adding plots as the data becomes available. For example, if a set of calculations done in a `for` loop generates vectors x and y at the end of each loop and you would like to plot them on the same graph, `hold` is the way to do it.

```
>> t=linspace(0,2*pi,100);          % Generate vector t
>> y1=sin(t);  y2=t;               % Calculate y1, y2, y3
>> y3=t-(t.^3)/6+(t.^5)/120;
>> plot(t,y1,t,y2,'--',t,y3,'o')    % Plot (t,y1) with solid line
                                   %- (t,y2) with dashed line and
                                   %- (t,y3) with circles
>> axis([0 5 -1 5])                % Zoom in with new axis limits
>> xlabel('t')                     % Put x-label
>> ylabel('Approximations of sin(t)')% Put y-label
>> title('Fun with sin(t)')        % Put title
>> text(3.5,0,'sin(t)')            % Write 'sin(t)' at point (3.5,0)
>> gtext('Linear approximation')
>> gtext('First 3 terms')
>> gtext('in Taylor series')
```

gtext writes the specified string at a location clicked with the mouse in the graphics window. So after hitting return at the end of gtext command, go to the graphics window and click a location.

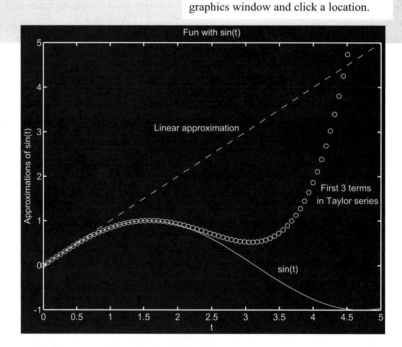

Figure 6.2: Example of an overlay plot along with examples of xlabel, ylabel, title, axis, text, and gtext commands. The three lines plotted are $y_1 = \sin t$, $y_2 = t$, and $y_3 = t - \frac{t^3}{3!} + \frac{t^5}{5!}$.

Method 3: Using the `line` command to generate overlay plots

The `line` is a low-level graphics command used by the `plot` command to generate lines. Once a plot exists in the graphics window, additional lines may be added by using the `line` command directly. The `line` command takes a pair of vectors (or a triplet in 3-D) followed by *parameter name/parameter value* pairs as arguments:

```
line(xdata, ydata, ParameterName, ParameterValue)
```

This command simply adds lines to the existing axes. For example, the overlay plot created by the previous script file could also be created with the following script file, which uses the `line` command instead of the `hold` command. As a bonus to the reader, we include an example of the `legend` command (see Section 6.1.2, page 177).

```
% -- Script file to generate an overlay plot with the line command --
% ------------------
% First, generate some data
t = linspace(0,2*pi,100);        % Generate vector t
y1 = sin(t);                     % Calculate y1, y2, y3
y2 = t;
y3 = t - (t.^3)/6 + (t.^5)/120;

% Now, plot the three lines
plot(t,y1)                       % Plot (t,y1) with (default) solid line
line(t,y2,'linestyle','--')      % Add line (t,y2) with dashed line and
line(t,y3,'marker','o',...       % Add line (t,y3) plotted with circles--
          'linestyle', 'none')   % but no line
% Adjust the axes
axis([0 5 -1 5])                 % Zoom in with new axis limits

% Dress up the graph
xlabel('t')                      % Put x-label
ylabel('Approximations of sin(t)')
                                 % Put y-label
title('Fun with sin(t)')         % Put title

legend('sin(t)','linear approx.','fifth-order approx.')
                                 % add legend
```

The output generated by the preceding script file is shown in Fig. 6.3. After generating the plot, click and hold the mouse on the legend rectangle and see if you can drag the legend to some other position. Alternatively, you could specify an option in the legend command to place the legend rectangle in any of the four corners of the plot. See the on-line help on `legend`.

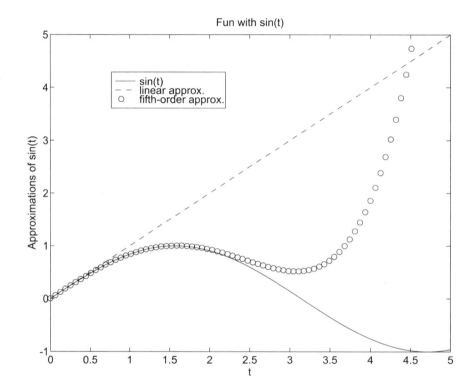

Figure 6.3: Example of an overlay plot produced by using the `line` command. The legend is produced by the `legend` command. See the script file for details.

6.1.6 Specialized 2-D plots

There are many specialized graphics functions for 2-D plotting. They are used as alternatives to the `plot` command we have just discussed. There is a whole suite of `ez` plotter functions, such as `ezplot, ezpolar, ezcontour`, that are truly easy to use. See Section 3.8 for a discussion and examples of these functions.

Here, we provide a list of other functions commonly used for plotting x-y data:

For on-line help type:
`help graph2d`
`help specgraph`

`area`	creates a filled area plot,
`bar`	creates a bar graph,
`barh`	creates a horizontal bar graph,
`comet`	makes an animated 2-D plot,
`compass`	creates arrow graph for complex numbers,
`contour`	makes contour plots,
`contourf`	makes filled contour plots,
`errorbar`	plots a graph and puts error bars,
`feather`	makes a feather plot,
`fill`	draws filled polygons of specified color,
`fplot`	plots a function of a single variable,
`hist`	makes histograms,
`loglog`	creates plot with log scale on both the x-axis and the y-axis,
`pareto`	makes pareto plots,
`pcolor`	makes pseudocolor plot of a matrix,
`pie`	creates a pie chart,
`plotyy`	makes a double y-axis plot,
`plotmatrix`	makes a scatter plot of a matrix,
`polar`	plots curves in polar coordinates,
`quiver`	plots vector fields,
`rose`	makes angled histograms,
`scatter`	creates a scatter plot,
`semilogx`	makes semilog plot with log scale on the x-axis,
`semilogy`	makes semilog plot with log scale on the y-axis,
`stairs`	plots a stair graph, and
`stem`	plots a stem graph.

On the following pages, we show examples of these functions. The commands shown in the middle column produce the plots shown in the right column. There are several ways you can use these graphics functions and many of them take optional arguments. The following examples should give you a basic idea of how to use these functions and what kind of plot to expect from them. For more information on any of these functions, see the on-line help.

Function	Example Script	Output
`fplot`	$f(t) = t \sin t, \;\; 0 \le t \le 10\pi$ `fplot('x.*sin(x)',[0 10*pi])` Note that the function to be plotted must be written as a function of x.	

`semilogx`	$x = e^{-t}, \;\; y = t, \; 0 \le t \le 2\pi$
	
	 `t = linspace(0,2*pi,200);`
	 `x = exp(-t); y = t;`
	 `semilogx(x,y), grid`	

`semilogy`	$x = t, \; y = e^{t}, \;\; 0 \le t \le 2\pi$
	
	 `t = linspace(0,2*pi,200);`
	 `semilogy(t,exp(t))`
	 `grid`	

`loglog`	$x = e^{t}, \; y = 100 + e^{2t}, \;\; 0 \le t \le 2\pi$
	
	 `t = linspace(0,2*pi,200);`
	 `x = exp(t);`
	 `y = 100 + exp(2*t);`
	 `loglog(x,y), grid`	

Function	Example Script	Output

polar	$r^2 = 2\sin 5t, \ \ 0 \le t \le 2\pi$
	
	 ```t = linspace(0,2*pi,200);```
	 ```r = sqrt(abs(2*sin(5*t)));```
	 ```polar(t,r)```	

fill	$\begin{aligned} r^2 &= 2\sin 5t, \ \ 0 \le t \le 2\pi \\ x &= r\cos t, \ \ y = r\sin t \end{aligned}$
	
	 ```t = linspace(0,2*pi,200);```
	 ```r = sqrt(abs(2*sin(5*t)));```
	 ```x = r.*cos(t);```
	 ```y = r.*sin(t);```
	 ```fill(x,y,'k'),```
	 ```axis('square')```	

bar	$\begin{aligned} r^2 &= 2\sin 5t, \ \ 0 \le t \le 2\pi \\ y &= r\sin t \end{aligned}$
	
	 ```t = linspace(0,2*pi,200);```
	 ```r = sqrt(abs(2*sin(5*t)));```
	 ```y = r.*sin(t);```
	 ```bar(t,y)```
	 ```axis([0 pi 0 inf]);```	

errorbar	$\begin{aligned} f_{\text{approx}} &= x - \frac{x^3}{3!}, \ \ 0 \le x \le 2 \\ error &= f_{\text{approx}} - \sin x \end{aligned}$
	
	 ```x = 0:.1:2;```
	 ```aprx2 = x - x.^3/6;```
	 ```er = aprx2 - sin(x);```
	 ```errorbar(x,aprx2,er)```	

Function	Example Script	Output

barh	World population by continents.
	
	``
	cont = char('Asia','Europe','Africa',...
	      'N. America','S. America');
	pop = [3332;696;694;437;307];
	barh(pop)
	for i=1:5,
	        gtext(cont(i,:));
	end
	xlabel('Population in millions')
	Title('World Population (1992)',
	      'fontsize',18)
	``	

plotyy	$$y_1 = e^{-x}\sin x,\ 0 \le t \le 10$$ $$y_2 = e^x$$
	
	``
	x = 1:.1:10;
	y1 = exp(-x).*sin(x);
	y2 = exp(x);
	Ax = plotyy(x,y1,x,y2);
	hy1 = get(Ax(1),'ylabel');
	hy2 = get(Ax(2),'ylabel');
	set(hy1,'string','e^-x sin(x)');
	set(hy2,'string','e^x ');
	``	

area	$$y = \frac{\sin(x)}{x},\quad -3\pi \le x \le 3\pi$$
	
	``
	x = linspace(-3*pi,3*pi,100);
	y = -sin(x)./x;
	area(x,y)
	xlabel('x'), ylabel('sin(x)./x')
	hold on
	x1 = x(46:55); y1 = y(46:55);
	area(x1,y1,'facecolor','y')
	``	

pie	World population by continents.
	
	``
	cont = char('Asia','Europe','Africa',...
	      'N. America','S. America');
	pop = [3332;696;694;437;307];
	pie(pop)
	for i=1:5,
	        gtext(cont(i,:));
	end
	Title('World Population (1992)',...
	      'fontsize',18)
	``	

Function	Example Script	Output

hist	Histogram of 50 randomly distributed numbers between 0 and 1.
	
	 ```y = randn(50,1);```
	 ```hist(y)```	

stem	$f = e^{-t/5}\sin t,\ 0 \le t \le 2\pi$
	
	 ```t = linspace(0,2*pi,200);```
	 ```f = exp(-.2*t).*sin(t);```
	 ```stem(t,f)```	

stairs	$$r^2 = 2\sin 5t,\ \ 0 \le t \le 2\pi$$ $$y = r\sin t$$ ```t = linspace(0,2*pi,200);```
	 ```r = sqrt(abs(2*sin(5*t)));```
	 ```y = r.*sin(t);```
	 ```stairs(t,y)```
	 ```axis([0 pi 0 inf]);```	

compass	$z = \cos\theta + i\sin\theta,\ \ -\pi \le \theta \le \pi$
	
	 ```th = -pi:pi/5:pi;```
	 ```zx = cos(th);```
	 ```zy = sin(th);```
	 ```z = zx + i*zy;```
	 ```compass(z)```	

Function	Example Script	Output
comet	$y = t \sin t, \ \ 0 \le t \le 10\pi$ ```\nq = linspace(0,10*pi,2000);\ny = q.*sin(q);\ncomet(q,y)\n``` (It is better to see it on screen.)	
contour	$z \ = \ -\dfrac{1}{2}x^2 + xy + y^2$ $\|x\| \le 5, \ \|y\| \le 5.$ ```\nr = -5:.2:5;\n[X,Y] = meshgrid(r,r);\nZ = -.5*X.^2 + X.*Y + Y.^2;\ncs = contour(X,Y,Z);\nclabel(cs)\n```	
quiver	$z \ = \ x^2 + y^2 - 5\sin(xy)$ $\|x\| \le 2, \ \|y\| \le 2.$ ```\nr = -2:.2:2;\n[X,Y] = meshgrid(r,r);\nZ = X.^2 - 5*sin(X.*Y) + Y.^2;\n[dx,dy] = gradient(Z,.2,.2);\nquiver(X,Y,dx,dy,2);\n```	
pcolor	$z \ = \ x^2 + y^2 - 5\sin(xy)$ $\|x\| \le 2, \ \|y\| \le 2.$ ```\nr = -2:.2:2;\n[X,Y] = meshgrid(r,r);\nZ = X.^2 - 5*sin(X.*Y) + Y.^2;\npcolor(Z), axis('off')\nshading interp\n```	

6.2 Using subplot for Multiple Graphs

If you want to make a few plots and place the plots side by side (not overlay), use the subplot command to design your layout. The subplot command requires three integer arguments:

$$\boxed{\texttt{subplot(m,n,p)}}$$

Subplot divides the graphics window into $m \times n$ subwindows and puts the plot generated by the next plotting command into the pth subwindow, where the subwindows are counted row-wise. Thus, the command subplot(2,2,3), plot(x,y) divides the graphics window into four subwindows and plots y versus x in the third subwindow, which is the first subwindow in the second row. For an example, see Fig. 6.5 on page 191.

6.3 3-D Plots

For on-line help type:
help graph3d

MATLAB provides extensive facilities for visualizing 3-D data. In fact, the built-in *colormaps* may be used to represent the fourth dimension. The facilities provided include built-in functions for plotting space curves, wire-frame objects, and shaded surfaces; generating contours automatically; displaying volumetric data; specifying light sources; interpolating colors and shading; and even displaying images. Typing **help graph3d** in the command window gives a list of functions available for general 3-D graphics. Here is a list of commonly used functions other than those from the ez-stable, such as ezsurf, ezmesh, ezplot3, discussed in Section 3.8.

plot3	plots curves in space,
stem3	creates discrete data plot with stems in 3-D,
bar3	plots 3-D bar graph,
bar3h	plots 3-D horizontal bar graph,
pie3	makes 3-D pie chart,
comet3	makes animated 3-D line plot,
fill3	draws filled 3-D polygons,
contour3	makes 3-D contour plots,
quiver3	draws vector fields in 3-D,
scatter3	makes scatter plots in 3-D,
mesh	draws 3-D mesh surfaces (wire-frame),
meshc	draws 3-D mesh surfaces along with contours,
meshz	draws 3-D mesh surfaces with reference plane curtains,
surf	creates 3-D surface plots,
surfc	creates 3-D surface plots along with contours,
surfl	creates 3-D surface plots with specified light source,
trimesh	mesh plot with triangles,
trisurf	surface plot with triangles,
slice	draws a volumetric surface with slices,
waterfall	creates a *waterfall* plot of 3-D data,

`cylinder`	generates a cylinder,
`ellipsoid`	generates an ellipsoid, and
`sphere`	generates a sphere.

Among these functions, `plot3` and `comet3` are the 3-D analogs of `plot` and `comet` commands mentioned in the 2-D graphics section. The general syntax for the `plot3` command is

$$\boxed{\texttt{plot3}(x,\ y,\ z,\ \texttt{'style-option'})}$$

This command plots a curve in 3-D space with the specified line style. The argument list can be repeated to make overlay plots, just the same way as with the `plot` command. A catalog of these functions with example scripts and the corresponding output is given on pages 196–200. Because the example scripts use a few functions that we discuss in Section 6.3.3, we postpone the catalog until then.

Plots in 3-D may be annotated with functions already mentioned for 2-D plots—`xlabel`, `ylabel`, `title`, `text`, `grid`, etc., along with the obvious addition of `zlabel`. The `grid` command in 3-D makes the 3-D appearance of the plots better, especially for curves in space (see Fig. 6.5 for example).

6.3.1 View

The viewing angle of the observer is specified by the command

$$\boxed{\texttt{view}(azimuth,\ elevation)}$$

where *azimuth* and *elevation* are angles specified in degrees. The azimuth is the rotation about the *z*-axis measured counterclockwise from the negative *y*-axis, and the elevation is the vertical angle measured positive above the *xy*-plane (see Fig. 6.4). The default values for these angles are $-37.5°$ and $30°$, respectively.

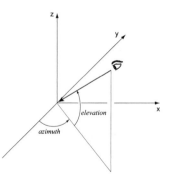

Figure 6.4: The viewing angles *azimuth* and *elevation* in 3-D plots.

By specifying appropriate values of the azimuth and the elevation, one can plot the projections of a 3-D object on different 2-D planes. For example, the command `view(90,0)` puts the viewer on the positive *x*-axis, looking straight on the *yz*-plane, and thus produces a 2-D projection of the object on the *yz*-plane. Figure 6.5 shows the projections obtained by specifying different view angles.

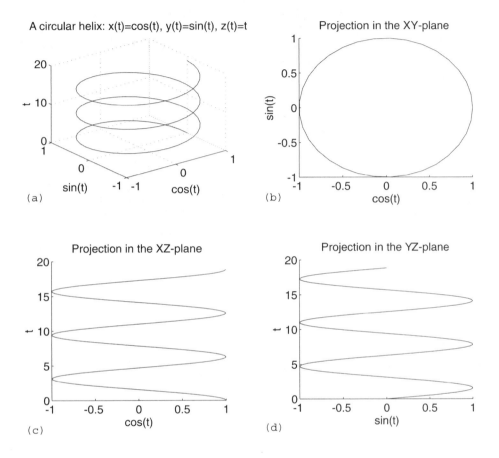

Figure 6.5: Examples of `plot3`, `subplot`, and `view`. Note how the 3-D grid in the background helps in the 3-D appearance of the space curve. The 3-D curve plotted here is generated with the commands `t=linspace(0,6*pi,100); x=cos(t); y=sin(t); z=t;` and then plotted as follows: (a) `subplot(2,2,1)`, `plot3(x,y,z)`, (b) `subplot(2,2,2)`, `plot3(x,y,z)`, `view(0,90)`, (c) `subplot(2,2,3)`, `plot3(x,y,z)`, `view(0,0)`, (d) `subplot(2,2,4)`, `plot3(x,y,z)`, `view(90,0)`. Labels and titles are added as discussed before. Although the three 2-D pictures could be made using the `plot` command, this example illustrates the use of viewing angles.

View(2) and view(3)

These are the special cases of the `view` command, specifying the default 2-D and 3-D views:

`view(2)`   same as `view(0,90)`, shows the projection in the xz-plane, and
`view(3)`   same as `view(-37.5,30)`, shows the default 3-D view.

The `view(3)` command can be used to see a 2-D object in 3-D. It may be useful in visualizing the perspectives of different geometrical shapes. The following script file draws a filled circle in 2-D and also views the same circle in 3-D. The output is shown in Fig. 6.6.

```
% ---- script file to draw a filled circle and view it in 3D ----
theta = linspace(0,2*pi,100);     % create vector theta
x = cos(theta);                   % generate x-coordinates
y = sin(theta);                   % generate y-coordinates
subplot(1,2,1)                    % initiate a 1 by 2 subplot
fill(x,y,'g'); axis('square');    % plot the filled circle
subplot(1,2,2)                    % go to the 2nd subplot
fill(x,y,'g'); axis('square');    % plot the same circle again
view(3)                           % view the 2-D circle in 3-D
```

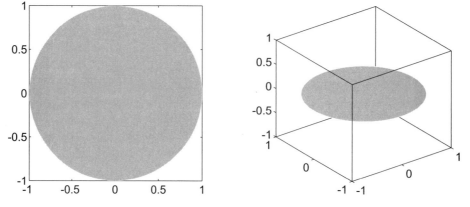

Figure 6.6: Example of `view(3)` to see a 2-D object in 3-D.

6.3.2   Rotate view

MATLAB provides more versatile and easier utilities for manipulating the view angle of a 3-D plot. Simply click on the *rotate in 3D* button located in the toolbar of the figure window (see Fig. 6.1 on page 178) and use your mouse to rotate the view. Alternatively, you can use a utility function called `rotate3d`. Simply turn it on with `rotate3d on` and rotate the view with your mouse. There are also some sophisticated *camera* functions that let you specify the camera view angle, zoom, roll, pan, etc. See the on-line help on `graph3D`.

6.3.3 Mesh and surface plots

The functions for plotting meshes and surfaces `mesh` and `surf`, and their various variants `meshz`, `meshc`, `surfc`, and `surfl`, take multiple optional arguments, the most basic form being `mesh(Z)` or `surf(Z)`, where Z represents a matrix. Usually surfaces are represented by the values of z-coordinates sampled on a grid of (x, y) values. Therefore, to create a surface plot we first need to generate a grid of (x, y) coordinates and find the height (z-coordinate) of the surface at each of the grid points. Note that you need to do the same thing for plotting a function of two variables. MATLAB provides a function `meshgrid` to create a grid of points over a specified range.

The function `meshgrid`

Suppose we want to plot a function $z = x^2 - y^2$ over the domain $0 \leq x \leq 4$ and $-4 \leq y \leq 4$. To do so, we first take several points in the domain, say 25 points, as shown in Fig. 6.7. We can create two matrices X and Y, each of size 5×5, and write the xy-coordinates of each point in these matrices. We can then evaluate z with the command `z=X.^2-Y.^2;`. Creating the two matrices X and Y is much easier with the `meshgrid` command:

```
rx = 0:4;              % create a vector rx=[0 1 2 3 4]
ry = -4:2:4;           % create a vector ry=[-4 -2 0 2 4]
[X,Y] = meshgrid(rx,ry);  % create a grid of 25 points and
                       %- store their coordinates in X and Y.
```

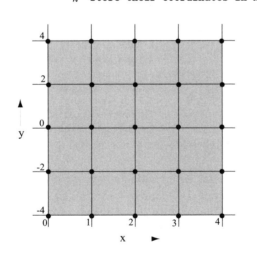

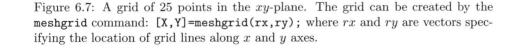

Figure 6.7: A grid of 25 points in the xy-plane. The grid can be created by the `meshgrid` command: `[X,Y]=meshgrid(rx,ry);` where rx and ry are vectors specifying the location of grid lines along x and y axes.

The preceding commands generate the 25 points shown in Fig. 6.7. All we need to generate is two vectors, rx and ry, to define the region of interest and distribution of grid points. Also, the two vectors need not be either same-sized or linearly spaced (although, most of the time, we take square regions and create grid points equally spaced in both directions; see examples on pages 196–200). To be comfortable with 3-D graphics, you should understand the use of `meshgrid`.

Back to `mesh` plot

When a surface is plotted with `mesh(z)` (or `surf(z)`) command, where z is a matrix, then the tick marks on the x-axis and the y-axis do not indicate the domain of z but the row and column indices of the z matrix. This is the default. Typing `mesh(x,y,z)` or `surf(x,y,z)`, where x and y are vectors used by `meshgrid` command to create a grid, results in the surface plot of z with x- and y-values shown on the x- and y-axes. The following script file should serve as an example of how to use `meshgrid` and `mesh` commands. Here we try to plot the surface

$$z = \frac{xy(x^2 - y^2)}{x^2 + y^2}, \quad -3 \le x \le 3, \ -3 \le y \le 3$$

by computing the values of z over a 50×50 grid on the specified domain. The results of the two plot commands are shown in Fig. 6.8.

```
%----------------------------------------------------------------------
% Script file to generate and plot the surface
% z =  xy(x^2-y^2)/(x^2+y^2) using meshgrid and mesh commands.
%----------------------------------------------------------------------
x = linspace(-3,3,50); y = x; % Generate 50 element long vectors x & y
[X,Y] = meshgrid(x,y);        % Create a grid over the specified domain
Z = X.*Y.*(X.^2-Y.^2)./(X.^2+Y.^2);  % Calculate Z at each grid point
mesh(X,Y,Z)                   % Make a wire-frame surface plot of Z and
                              %- use x and y values on the x and y-axes
title('Plot created by mesh')
figure(2)                     % Open a new figure window
meshc(X,Y,Z),view(-55,20)     % Plot the same surface along with
                              %- contours and show the view from
                              %- the specified angles
title('Plot created by meshc')
```

Surfaces created by `mesh` or its variants have a wire-frame appearance, whereas surfaces created by the **surf** command or its variants produce a true surface-like appearance, especially when used with the **shading** command. There are three kinds of shading available—**shading flat** produces simple flat shading, **shading interp** produces more dramatic interpolated shading, and **shading faceted**, the default shading, shows shaded facets of the surface. Both `mesh` and `surf` can plot parametric surfaces with color scaling to indicate a fourth dimension. This is accomplished by giving four matrix arguments to these commands, e.g., `surf(X,Y,Z,C)` where X, Y, and Z are matrices representing a surface in parametric form and C is the matrix indicating color scaling. The command `surfl` can be used to control the light reflectance and to produce special effects with a specified location of a light source. See on-line help on `surfl` for more information.

Plot created by mesh

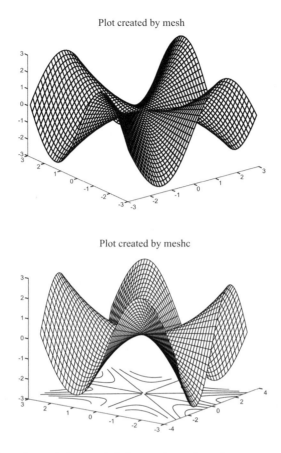

Plot created by meshc

Figure 6.8: 3-D surface plots created by `mesh` and `meshc` commands. The second plot uses a different viewing angle to show the center of the contour lines. Note that the surfaces do not show hidden lines (this is the default setting; it can be changed with the `hidden` command).

We close this section with a catalog of popular 3-D graphics functions on the following pages. We hope that you can use these functions for your needs simply by following the example scripts. But we acknowledge that the `meshgrid` command takes some thought to understand well.

Function	Example Script	Output
plot3	Plot of a parametric space curve: $$x(t) = t, \ y(t) = t^2, \ z(t) = t^3.$$ $$0 \le t \le 1.$$ ```t = linspace(0,1,100); x = t; y = t.^2; z = t.^3; plot3(x,y,z), grid xlabel('x(t) = t') ylabel('y(t) = t2') zlabel('z(t) = t3')```	
fill3	Plot of four filled polygons with three vertices each. ```X = [0 0 0 0; 1 1 -1 1; 1 -1 -1 -1]; Y = [0 0 0 0; 4 4 4 4; 4 4 4 4]; Z = [0 0 0 0; 1 1 -1 -1; -1 1 1 -1]; fillcolor=rand(3,4); fill3(X,Y,Z,fillcolor) view(120,30)```	
contour3	Plot of 3-D contour lines of $$z = -\frac{5}{1 + x^2 + y^2},$$ $$|x| \le 3, |y| \le 3.$$ ```r = linspace(-3,3,50); [x,y] = meshgrid(r,r); z = -5./(1 + x.^2 + y.^2); contour3(x,y,z)```	

Function	Example Script	Output

surf	$z = \cos x \cos y \, e^{\frac{-\sqrt{x^2+y^2}}{4}}$
	
	 $|x| \le 5, \quad |y| \le 5$
	
	 ```u = -5:.2:5;```
	```[X,Y] = meshgrid(u, u);```
	```Z = cos(X).*cos(Y).*...```
	``` exp(-sqrt(X.^2 + Y.^2)/4);```
	```surf(X,Y,Z)```	

surfc	$z = \cos x \cos y \, e^{\frac{-\sqrt{x^2+y^2}}{4}}$
	
	 $|x| \le 5, \quad |y| \le 5$
	
	 ```u = -5:.2:5;```
	```[X,Y] = meshgrid(u, u);```
	```Z = cos(X).*cos(Y).*...```
	``` exp(-sqrt(X.^2 + Y.^2)/4);```
	```surfc(Z)```
	```view(-37.5,20)```
	```axis('off')```	

surfl	$z = \cos x \cos y \, e^{\frac{-\sqrt{x^2+y^2}}{4}}$
	
	 $|x| \le 5, \quad |y| \le 5$
	
	 ```u = -5:.2:5;```
	```[X,Y] = meshgrid(u, u);```
	```Z = cos(X).*cos(Y).*...```
	``` exp(-sqrt(X.^2 + Y.^2)/4);```
	```surfl(Z)```
	```shading interp```
	```colormap hot```	

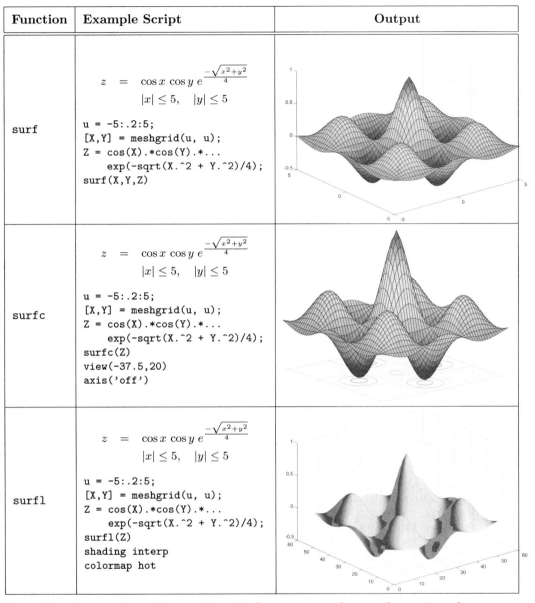

Note: Plotting a surface with `surf(X,Y,Z)` shows proper values on the x-axis and the y-axis, whereas plotting the surface with `surf(Z)` shows the row and column indices of matrix Z on the x-axis and the y-axis. The same is true for other 3-D plotting commands such as `mesh` and `contour3`. Compare the values on the x-axis and the y-axis in the first and the last figure in this table.

Function	Example Script	Output
mesh	$$z \;=\; -\frac{5}{1+x^2+y^2}$$ $$|x| \le 3, \quad |y| \le 3$$ ```x = linspace(-3,3,50);
y = x;
[x,y] = meshgrid(x,y);
z = -5./(1+x.^2+y.^2);
mesh(z)```	
meshz	$$z \;=\; \sin^2 x + \sin^2 y$$ $$|x| \le \frac{\pi}{2}, \quad |y| \le \frac{\pi}{2}$$ ```x = linspace(-pi/2,pi/2,50);
y = x;
[x,y] = meshgrid(x,y);
z = sin(x.^2) + sin(y.^2);
meshz(x,y,z), axis tight
view(-37.5, 50)```	
waterfall	$$z \;=\; -\frac{5}{1+x^2+y^2}$$ $$|x| \le 3, \quad |y| \le 3$$ ```x = linspace(-3,3,50);
y = x;
[x,y] = meshgrid(x,y);
z = -5./(1 + x.^2 + y.^2);
waterfall(z)
hidden off```	

Function	Example Script	Output
pie3	World population by continents. ```\n% popdata: Af,As,Eu,NA,SA\npop = [807;3701;731;481;349];\ncontinents = {'Africa','Asia',...\n'Europe','N.America','S.America'};\npie3(pop,continents)\nTitle({'World Population',...\n'(2003)'})\n```	 World Population (2003)
stem3	Discrete data plot with stems. $$x = t, \quad y = t\sin(t),$$ $$z = e^{t/10} - 1$$ for $0 \le t \le 6\pi$. ```\nt = linspace(0,6*pi,200);\nx = t; y = t.*sin(t);\nz = exp(t/10)-1;\nstem3(x,y,z,'filled')\n```	
ribbon	2-D curves as ribbons in 3-D. $$y_1 = \sin(t), \quad y_2 = e^{-.15t}\sin(t)$$ $$y_3 = e^{-.8t}\sin(t)$$ for $0 \le t \le 5\pi$. ```\nt = linspace(0,5*pi,100);\ny1 = sin(t);\ny2 = exp(-.15*t).*sin(t);\ny3 = exp(-.8*t).*sin(t);\ny = [y1; y2; y3];\nrib_width = 0.2;\nribbon(t',y',rib_width)\n```	

Function	Example Script	Output
| sphere | A unit sphere centered at the origin and generated by three matrices $x$, $y$, and $z$ of size $21 \times 21$ each.<br><br>```
sphere(20)
axis('square')
```<br>or<br>```
[x,y,z] = sphere(20);
surf(x,y,z)
axis('square')
``` | |
| ellipsoid | An ellipsoid of radii $rx = 1$, $ry = 2$, and $rz = 0.5$, centered at the origin.<br><br>```
cx = 0; cy = 0; cz = 0;
rx = 1; ry = 2; rz = 0.5;
ellipsoid(cx,cy,cz,rx,ry,rz)
axis('equal')
``` | |
| cylinder | A cylinder generated by<br><br>$$r = \sin(3\pi z) + 2$$<br>$$0 \le z \le 1, \quad 0 \le \theta \le 2\pi.$$<br><br>```
z = [0:.02:1]';
r = sin(3*pi*z) + 2;
cylinder(r), axis square
``` | |
| slice | Slices of the volumetric function $f(x,y,z) = \cos^2 x + \cos^2 y - z^2$ $|x| \le 3$, $|y| \le 3$, $|z| \le 3$ at $x = -2$ and 2, $y = 2$, and $z = -2.5$ and 0.<br><br>```
v = [-3:.2:3];
[x,y,z] = meshgrid(v,v,v);
f = (cos(x).^2 + sin(y).^2-z.^2);
xv = [-2 2.5]; yv = 2;
zv = [-2.5 0];
slice(x,y,z,f,xv,yv,zv);
```<br><br>The value of the function is indicated by the color intensity. | |
```

6.3.4   Vector field and volumetric plots

One of the most crucial needs of visualization in scientific computation is for data that is essentially volumetric, i.e., defined over a 3-D space. For example, we may have temperature or pressure defined over each (x, y, z) triple in a bounded 3-D space. How do we display this data graphically? If we have a function $z = f(x, y)$ defined over a finite region of the xy-plane, we can display z or f as a 3-D surface. But we have $f(x, y, z)$! So, we need a 4-D hypersurface. That is the basic problem.

For on-line help type:
`help vissuite`

We display volumetric data by slicing it along several planes in 3-D and plotting the data on those planes, either with graded color maps or with contours. Such displays are still an area of active research. However, we can do fairly well with the tools currently available. One of the most common applications is in the area of 3-D vector fields. A vector field defines a vector quantity as a function of the space variables (x, y, z). Fortunately, in this case we can display the data (the vector) with an arrow drawn at each (x, y, z) triple, with the magnitude of the vector represented by the length of the arrow, and the direction represented by the orientation of the arrow. This concept is used extensively in dynamical systems in various ways.

MATLAB provides extensive tools for visualizing vector fields and volumetric data. Unfortunately, these tools are beyond the scope of this book. Therefore, we merely mention the tools here and give examples of only the "most likely to be used" tools.

Plotting vector fields

The plotting functions available in MATLAB for vector field visualization include `quiver`, `quiver3`, `stream2`, `stream3`, `streamline`, `streamtube`, `streamribbon`, `streamslice`, `streamparticles`, `coneplot`, `divergence`, `curl`, etc.

If $u(x, y)$ and $v(x, y)$ are given as vector components in x- and y-directions, respectively, then the vector field can be easily drawn with `quiver` (`quiver3` in 3-D). An example of `quiver` appears on page 188 in the table of 2-D plots. The *stream* functions are an extension of the same concept; they draw streamlines or trajectories from user-specified points in the specified vector field. This suite of functions has been a welcome addition in MATLAB (version 6 onward).

The function `streamline` is useful for drawing solution trajectories in 2-D and 3-D vector fields defined by ODEs. You need not solve the ODEs!
Example: Let

$$\begin{aligned}\dot{x} &= y + x - x(x^2 + y^2) \quad \text{and} \\ \dot{y} &= -x + y - y(x^2 + y^2).\end{aligned}$$

These two ODEs define a vector field ($u \equiv \dot{x}$ and $v \equiv \dot{y}$). Let us use `streamline` to draw a few solution trajectories starting from various points in the xy-plane (initial conditions in the phase plane). The general syntax of `streamline` is

```
streamline(x,y,z, u,v,w, x0,y0,z0)
```

where (x, y, z) are 3-D matrices of grid points where the vector field components (u, v, w) are specified, and $(x0, y0, z0)$ are starting points for the streamlines to be drawn. Here is a script file that draws the streamlines for our 2-D vector field.

```
% STREAMLINE2D example of using streamline for 2-D vector field
% The vector field is given by two ODEs
% -----------------------------------
  % create grid points in 2-D
    v = linspace(-2,2,50);
    [X,Y] = meshgrid(v);
  % define vector field
    U = Y + X - X.*(X.^2 + Y.^2);
    V = Y - X - Y.*(X.^2 + Y.^2);
  % specify starting points for streamlines
    x0 = [-2 -2 -2 -2 -.5 -.5 .5 .5 2 2 2 2 -.01 -.01 .01 .01];
    y0 = [-2 -.5 .5 2 -2 2 -2 2 -2 -.5 .5 2 -.01 .01 -.01 .01];
  % draw streamlines
    streamline(X,Y,U,V,x0,y0)
    axis square
```

The result obtained is shown in Fig. 6.9.

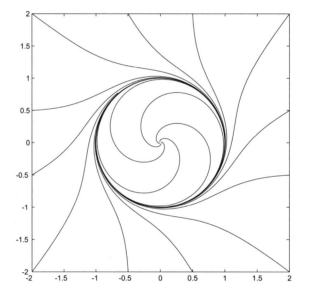

Figure 6.9: Plot obtained by executing the script file `streamline2D`.

Plotting volumetric data

The functions available in MATLAB for volumetric data visualization include `slice`, `slicecontour`, `isosurface`, `isonormal`, `isocaps`, `isocolors`, `subvolume`, `reducevolume`, `smooth3`, `reducepath`, etc. See on-line help for more details.

6.3.5 Interpolated surface plots

Many times, we get data (usually from experiments) in the form of (x, y, z) triples and we want to fit a surface through the data. Thus, we have a vector z that contains the z-values corresponding to irregularly spaced x- and y-values. Here, we do not have a regular grid, as created by `meshgrid`, and we do not have a matrix Z that contains the z-values of the surface at those grid points. Therefore, we have to fit a surface through the given triplets (x_i, y_i, z_i). The task is much simpler than it seems. MATLAB provides a function, `griddata`, that does this interpolation for us. The general syntax of this function is

```
[Xi,Yi,Zi] = griddata(x,y,z,xi,yi,method)
```

where x, y, z are the given data vectors (nonuniformly spaced), xi and yi are the user-prescribed points (hopefully, uniformly spaced) at which zi are computed by interpolation, and *method* is the choice for the interpolation algorithms. The algorithms available are *nearest*, *linear*, *cubic*, and *v4*. See the on-line documentation for a description of these methods.

As an example, let us consider 50 randomly distributed points in the xy-plane, in the range $-1 < x < 1$ and $-1 < y < 1$. Let the z-values at these points be given by $z = 3/(1 + x^2 + y^2)$. Thus, we have three vectors of length 50 each. The data points are shown in Fig. 6.10 using the `scatter3` plot. Now, let us fit a surface through these points:

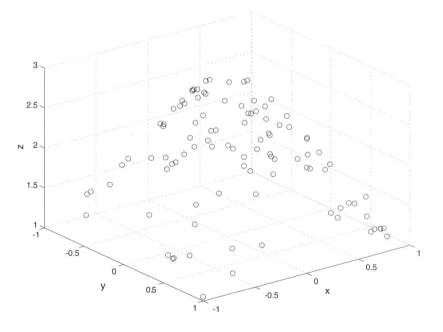

Figure 6.10: Nonuniformly distributed data points (x, y, z).

```
% SURFINTERP: script file to generate an interpolated surface
% Given vectors x, y, and z, generate data matrix Zi from
% interpolation to fit a surface through the data
% -------------------------------------------------------------
xv = 2*rand(1,100)-1;        % this is the given x
yv = 2*rand(1,100)-1;        % this is the given y
zv = 3./(1 + xv.^2 + yv.^2); % this is the given z
scatter3(xv,yv,zv)           % show data as stem plot

xi = linspace(-1,1,30);      % create uniformly spaced xi
yi = xi';                    % create uniformly spaced yi
                             % note that yi is a column

[Xi,Yi,Zi] = griddata(xv,yv,zv,xi,yi,'v4');
                             % interpolate surface using
                             % v4 (MATLAB 4 griddata) method
surf(Xi,Yi,Zi)              % plot the interpolated surface
```

The interpolated surface is shown in Fig. 6.11.

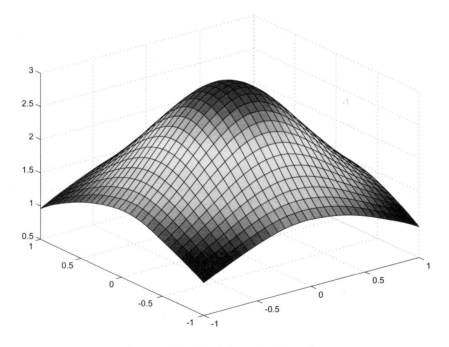

Figure 6.11: The interpolated surface.

6.4   Handle Graphics

For on-line help
type:
`help graphics`

> You need not learn or understand Handle Graph-
> ics to do most of the plotting an average person
> needs. If you want extra-detailed control of your
> graph appearance or want to do animation (be-
> yond `comet` plots), you might want to learn Han-
> dle Graphics. This is NOT a topic for beginners.

A line is a graphics object. It has several properties—line style, color, thickness, visibility, etc. Once a line is drawn on the graphics screen, it is possible to change any of its properties later. Suppose you draw several lines with pencil on a paper. If you want to change one of the lines, you must first find the line you want to change and then change what you do not like about it. On the graphics screen, a line may be one among several graphics objects (e.g., axes, text, labels). So how do you get hold of a line? You get hold of a line by its *handle*.

What is a handle? MATLAB assigns a floating-point number to every object in the figure window (including invisible objects), and it uses this number as an address or name for the object in the figure. This number is the handle of the object.

Once you get hold of the handle, you can access all properties of the object. In short, the handle identifies the object and the object brings with it the list of its properties. In programming, this approach of defining objects and their properties is called *object-oriented programming*. The advantage it offers is that you can access individual objects and their properties and change any property of an object without affecting other properties or objects. Thus, you get complete control over graphics objects. MATLAB's entire system of object-oriented graphics and its user controlability is Handle Graphics. Here, we briefly discuss this system and its usage, but we urge the more interested reader to consult *Using MATLAB Graphics* [6] for more details.

The key to understanding and using the Handle Graphics system is to know how to get the handles of graphics objects and how to use handles to get and change properties of the objects. Not all graphics objects are independent (for example, the appearance of a line depends on the current axes in use), and a certain property of one may affect the properties of the others. It is, therefore, important to know how the objects are related.

6.4.1   The object hierarchy

Graphics objects follow a hierarchy of parent-child relationship. The following tree diagram shows the hierarchy.

It is important to know this structure for two reasons:

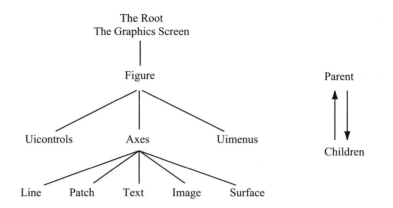

1. It shows you which objects will be affected if you change a default property value at a particular level.

2. It tells you at which level you can query for handles of which objects.

6.4.2 Object handles

Object handles are unique identifiers associated with each graphics object. These handles have a floating-point representation. Handles are created at the time of creation of the object by graphics functions such as `plot(x,y)`, `contour(z)`, `line(z1,z2)`, and `text(xc,yc,'Look at this')`.

Getting object handles

There are two ways of getting hold of handles:

1. By creating handles explicitly at the object-creation-level commands (that is, you can make a plot and get its handle at the same time):

   `hl = plot(x,y,'r-')`   returns the handle of the line to *hl*, and
   `hxl = xlabel('Angle')`   returns the handle of the *x*-label to *hxl*.

2. By using explicit handle-returning functions:

`gcf`	gets the handle of the current figure.
	Example: `hfig = gcf;` returns the handle of the current figure in *hfig*.
`gca`	gets the handle of the current axes.
	Example: `haxes = gca;` returns the handle of the current axes in *haxes*.
`gco`	gets the handle of the current object.

Handles of other objects, in turn, can be obtained with the **get** command. For example, `hlines=get(gca,'children')` returns the handles of all *children* of the current axes in a column vector `hlines`. The function **get** is used to get a property value of an object, specified by its handle, in the following command form:

get(*handle,'PropertyName'*).

For an object with handle h, type `get(h)` to get a list of all property names and their current values.

Examples:

`h1 = plot(x,y)`	plots a line and returns the handle hl of the plotted line,
`get(h1)`	lists all properties of the line and their values,
`get(h1,'type')`	shows the type of the object (e.g., line), and
`get(h1,'linestyle')`	returns the current line style of the line.

For more information on `get`, see the on-line help.

6.4.3 Object properties

Every graphics object on the screen has certain properties associated with it. For example, the properties of a line include `type`, `parent`, `visible`, `color`, `linestyle`, `linewidth`, `xdata`, `ydata`, etc. Similarly, the properties of a text object, such as `xlabel` or `title`, include `type`, `parent`, `visible`, `color`, `fontname`, `fontsize`, `fontweight`, `string`, etc. Once the handle of an object is known, you can see the list of its properties and their current values with the command `get(handle)`. For example, see Fig. 6.12 for the properties of a line and their current values.

There are some properties common to all graphics objects. These properties are `children`, `clipping`, `parent`, `type`, `userdata`, and `visible`.

Setting property values

You can see the list of properties and their values with the command `set(handle)`. Any property can be changed by the command

set(*handle, 'PropertyName', 'PropertyValue'*)

where *PropertyValue* may be a character string or a number. If *PropertyValue* is a string, then it must be enclosed within single quotes.

Figure 6.12 shows the properties and property values of a line.

Now let us look at an example.

Example: We create a line along with an explicit handle and then use the `set` command to change the line style, its thickness, and some of the data. See page 209.

```
>> t = linspace(0,pi,50);
>> hL = line(t,sin(t));
```
Create a line with handle *hL*.

```
>> get(hL);
  Color = [0 0 1]
  EraseMode = normal
  LineStyle = -
  LineWidth = [0.5]
  Marker = none
  MarkerSize = [6]
  MarkerEdgeColor = auto
  MarkerFaceColor = none
  XData = [ (1 by 50) double array]
  YData = [ (1 by 50) double array]
  .
  .
  .
```
Query the line's properties and their current values with the get command.

(many more properties follow)

```
>> set(hL)
  Color
  EraseMode: [ {normal} | background | xor | none ]
  LineStyle: [ {-} | -- | : | -. | none ]
  LineWidth
  Marker: [ + | o | * | . | x | square | diamond ..
  MarkerSize
  MarkerEdgeColor: [ none | {auto} ] -or- a ColorSpec.
  MarkerFaceColor: [ {none} | auto ] -or- a ColorSpec.
  XData
  YData
  ZData
  .
  .
  .
```
Query the line's properties that can be set and the available options.

(many more properties follow)

Figure 6.12: Example of creating a line with an explicit handle and finding the properties of the line, along with their current values.

Example Script	Output

Create a simple line plot and assign its handle to *hL*.

```
t = linspace(0,pi,50);
x = t.*sin(t);
hL = line(t,x);
```

Change the line style to dashed.

```
set(hL,'linestyle','--')
```

Change the line thickness.

```
set(hL,'linewidth',3,'marker','o')
```

Change the values of some *y*-coordinates by changing data points.

```
yvec = get(hL,'ydata');
yvec(15:20) = 0;
yvec(40:45) = 0;
set(hL,'ydata',yvec)
```

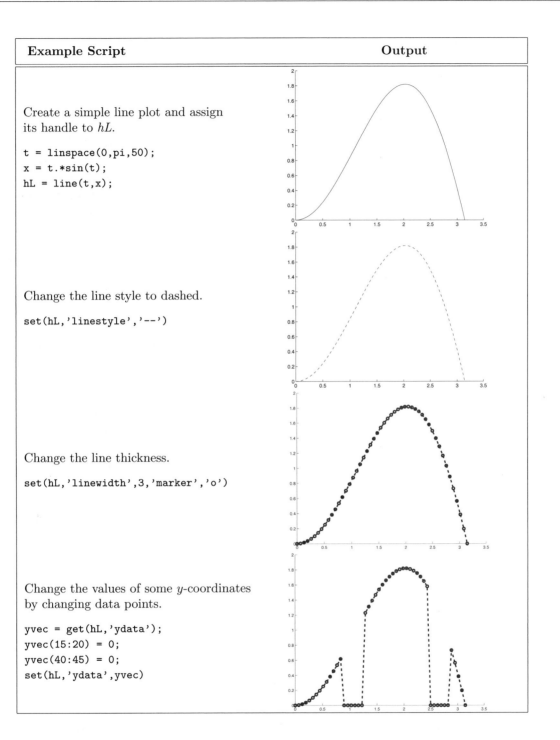

6.4.4 Modifying an existing plot

Even if you create a plot without explicitly creating object handles, MATLAB creates handles for each object on the plot. If you want to modify any object, you have to first get its handle. Here is where you need to know the parent-child relationship among several graphics objects. The following example illustrates how to get the handle of different objects and use the handles to modify the plot.

We take Fig. 6.2 on page 180 and use the aforementioned Handle Graphics features to modify the figure. The following script file is used to change the plot in Fig. 6.2 to the one shown in Fig. 6.13. You may find the following script file confusing because it uses a vector of handles, *hline*, and accesses different elements of this vector, without much explanation. Hopefully, your confusion will be cleared after you read the section on understanding a vector of handles. [Note: Before using the following commands, you must execute the commands shown in Fig. 6.2.]

```
h = gca;                              % get the handle of the current axes
set(h,'box','off');                   % throw away the enclosing box frame
hline = get(h,'children');            % get the handles of children of axes
                                      %- Note that hline is a vector of
                                      %- handles because h has many  children
set(hline(7),'linewidth',4)           % change the line width of the 1st  line
set(hline(6),'visible','off')         % make the 'lin. approx' line  invisible
delete(hline(3))                      % delete the text 'linear  approximation'
hxl = get(h,'xlabel');                % get the handle of xlabel
set(hxl,'string','t (angle)')         % change the text of xlabel
set(hxl,'fontname','times')           % change the font of xlabel
set(hxl,'fontsize',20,'fontweight','bold')
                                      % change the font-size & font-weight
```

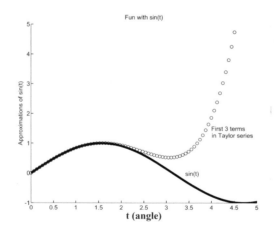

Figure 6.13: Example of manipulation of a figure with Handle Graphics. This figure is a result of executing the preceding script file after generating Fig. 6.2.

Understanding a vector of handles

In the previous script file, you may perhaps be confused about the use of the handle *hline*. The command, `hline=get(h,'children')`, gets the handles of all the children of the current axes (specified by handle *h*) in a column vector *hline*. The vector *hline* has seven elements—three handles for the three lines and four handles for the four text objects (created by `text` and `gtext` commands). So, how do we know which handle is for which line or which text? The command `get(hline(i),'type')` lists the type of the object whose handle is *hline(i)*. The confusion is not clear yet. What if *hline*(5), *hline*(6), and *hline*(7) are all lines? How do we know which handle corresponds to which line? Once we know the type of the object, we can identify its handle, among several similar object handles, by querying a more distinctive property of the object, such as `linestyle` for lines and `string` for text objects. For example, consider the handle vector *hline* given earlier. Then,

`get(hline(5),'marker')`	returns o for the line style,
`get(hline(6),'linestyle')`	returns -- for the line style,
`get(hline(1),'string')`	returns in Taylor series for the string, and
`get(hline(2),'string')`	returns First 3 terms for the string.

From this example, it should be clear that *the handles of children of the axes are listed in the stacking order of the objects*, i.e., the last object added goes on the top of the stack. Thus, the elements of the handle vector correspond to the objects in the reverse order of their creation!

Deleting graphics objects

Any object in the graphics window can be deleted without disturbing the other objects with the command

`delete(ObjHandle)`

where *ObjHandle* is the handle of the object. We have used this command in the script file that produced Fig. 6.13 to delete the text "linear approximation" from the figure. We could have used `delete(hline(6))` to delete the corresponding line rather than making it invisible.

Modifying plots with PropEdit

Now that you have some understanding of Handle Graphics, object handles, and object properties, you may like to use the point-and-click graphics editor, *PropEdit*. Simply type `propedit` to activate the editor. All graphics objects from the active figure window are shown, along with their properties, in the PropEdit window. You can select a property from the list by clicking on it and then change it in the narrow rectangle in the middle. A graphics object with a plus (+) on its left indicates that you can double-click on it to see its children.

6.4.5   Complete control over the graphics layout

We close this section with an example of arbitrary placement of axes and figures in
the graphics window. With Handle Graphics tools such as these, you have almost
complete control of the graphics layout. Here are two examples.

Example 1: Placing insets

The following script file shows how to create multiple axes, size them, and place
them so that they look like insets. The output appears in Fig. 6.14.

```
% INSETGRAPHICS:  Example script for creating insets in plots
%-------------------------------------------------------------------
%          Example of graphics placement with Handle Graphics
%-------------------------------------------------------------------

clf                                    % clear figure window
t = linspace(0,2*pi);  t(1)=eps;       % t(1) is set to a small number
y = sin(t);
%--------------------

h1 = axes('position',[0.1 0.1 .8 .8]); % place axes with width .8 and
                                       %- height .8 at coordinates (.1,.1)
plot(t,y),xlabel('t'),ylabel('sin t')
set(h1,'Box','Off');                   % Turn the enclosing box off
xhl = get(gca,'xlabel');               % get the handle of 'xlabel' of the
                                       %- current axes and assign to  xhl
set(xhl,'fontsize',16,'fontweight','bold')
                                       % change attributes of 'xlabel'
yhl = get(gca,'ylabel');               % do the same with 'ylabel'
set(yhl,'fontsize',16,'fontweight','bold')

h2 = axes('position',[0.6 0.6 .2 .2]);% place another axes on the same plot
fill(t,y.^2,'r')                       % draw a filled polygon with red fill
set(h2,'Box','Off');
xlabel('t'),ylabel('(sin t)^2')
set(get(h2,'xlabel'),'FontName','Times')
set(get(h2,'ylabel'),'FontName','Times')

h3 = axes('position',[0.15 0.2 .3 .3]); % place yet another axes
polar(t,y./t);                         % make a polar plot
polarch = get(gca,'children');         % get the handle of all children
                                       %- of the current axes
set(polarch(1),'linewidth',3)          % set the line width of the first child
                                       %- which is the line we plotted

%----------------------
```

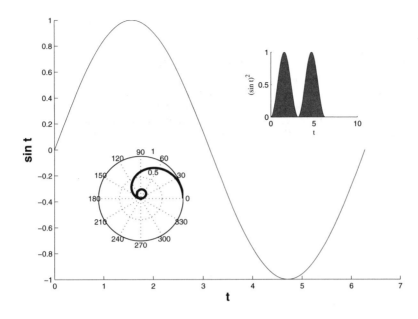

Figure 6.14: Example of manipulation of the figure window with Handle Graphics. Virtually anything in the figure window, including the placement of axes, can be manipulated with Handle Graphics.

Example 2: Fun with spirals

Now that you know how to create axes, position them, and make plots in them, let us have some fun. How about doing some artwork with MATLAB? Place and size the axes, take the outer box off at will, change the color of the axes, put the x-axis on top, y-axis on the right, and so on. Let us create a spiral given by

$$r(\theta) = e^{-\frac{\theta}{10}}, \quad 0 \le \theta \le 8\pi$$

and plot it as a filled spiral (with different color-fills) in four differently sized axes.

Go ahead, try the following script file. It produces the spirals that appear in Fig. 6.15 (you will, of course, see the spirals in color on your screen). Note how the same data produce progressively smaller spirals because of the size of the different axes.

```
% FUNWITHSPIRALS: Script to plot 4 filled spirals in different axes
% Written by Rudra Pratap on July 7, 1997,
%            last modified Nov 7, 2004.
% -------------------------
t = linspace(0,8*pi,200);   % create basic data for a spiral
r = exp(-.1*t);
x = r.*cos(t);
y = r.*sin(t);
```

```
clf                          % clear previous figure settings
h1 = axes('position',[.1,.1,.5,.5]);      % first axes
fill(x,y,'g')                % first (big) spiral in green
h2 = axes('position',[.45,.45,.3,.3]);    % second axes
fill(x,y,'b')                % second spiral in blue
set(h2,'xcolor','b');        % change x-axis color to blue
set(h2,'ycolor','b');        % change y-axis color to blue
set(h2,'xticklabel',' ');    % remove axis tick marks
set(h2,'yticklabel',' ');
h3 = axes('position',[.67,.67,.2,.2]);    % third axes
fill(x,y,'m'), box('off')    % third spiral, no outer box
set(h3,'xcolor','m');        % change axis color to magenta
set(h3,'ycolor','m');
set(h3,'xticklabel',' ');    % remove axis tick marks
set(h3,'yticklabel',' ');
h4 = axes('position',[.84,.84,.1,.1]);    % fourth axes
fill(x,y,'r')                % fourth spiral in red
set(h4,'color','y');         % change background color
set(h4,'xaxisloc','top');    % locate x-axis on top
set(h4,'yaxisloc','right'); % locate y-axis on right
```

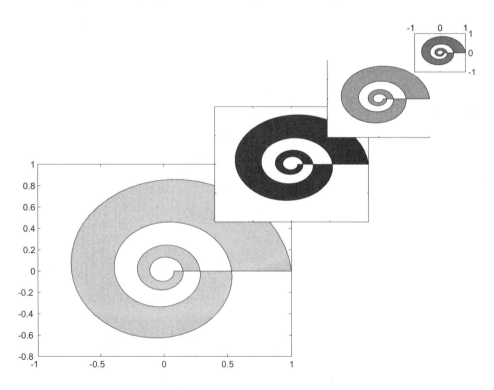

Figure 6.15: Example of manipulation of axes and its various properties.

6.5 Fun with 3-D Surface Graphics

6.5.1 Generating a cylindrical surface

There are specialized graphics functions—`cylinder`, `sphere`, and `ellipsoid`—for generating cylindrical, spherical, and ellipsoidal surfaces very easily. We can generate these surfaces even without using these specialized functions, using their mathematical equations. These functions make the task easier by reducing the number of steps required in generating these surfaces. Here, we use the function `cylinder` to create some interesting surfaces that also appear on the cover page of this book.

A cylinder is generated by defining a *generating curve* and revolving it around the z-axis by one complete revolution to sweep the desired cylindrical surface. The simplest case is a straight line (parallel to the z-axis) that generates a constant radius cylinder. The generating curve can be created by defining $r = f(z)$. For example, let

$$r(z) = r_0 + \sin(3\pi z), \qquad 0 \le z \le 1.$$

We take $r_0 = 1.5$ and create a cylinder with this generating curve using the following simple code.

```
z = linspace(0,1,101);      % take 101 points between 0 and 1
r = sin(3*pi*z)+1.5;        % define the generating curve r(z)
[X,Y,Z] = cylinder(r,50);   % generate surface data for the cylinder
```

The `cylinder` function automatically generates the appropriate matrices X, Y, and Z that can be used as arguments of the `surf` function to plot the cylinder. So, now we use the data to plot the cylinder with the following commands:

```
c = surf(X,Y,Z);        % plot the cylinder; its handle is c
axis square;            % set the axes to be square
view(-38.5,26);         % change the viewing angle slightly
```

The resulting plot is shown in Fig. 6.16(a). Now, let us spruce up this graph just a bit by changing its patch color, edge color, and transparency, and add a light source to enhance the 3-D look.

6.5.2 Face color, transparency, and light reflection

It is amazingly simple to manipulate 3-D graphics in MATLAB. Once you have a surface plot on the screen, you can activate the interactive plot editor by clicking on the white arrow button on the menu bar of the figure window. Now double-click on the object you want to edit and click on the **More Properties** button in the **Property Editor** subwindow. You can now select and change whichever property of the graph you like, including color of patch faces, color of edges, transparencies (alpha) of these objects, light sources, and color schemes. We, however, recommend that you familiarize yourself with how to change these properties using graphics handles. It is then easy to record such changes in your code and save the code for later use.

Surface plotting functions, `surf`, `surfc`, and `surfl`, use *patches* to create the surface. A `patch` is a 2-D graphics primitive (2-D equivalent of `line`) that is defined by its edges and its face. Both the *edge* and the *face* can take any color, set by specifying an RGB (red, green, blue) triplet or a predefined color name, and have independent transparencies, set by their respective *alpha* values (a `FaceAlpha` value of 1 makes the face opaque and a value of 0 makes it completely transparent).

One can also add a light source at a location of one's choice, specify its color and style. The most useful *style* perhaps is `'infinite'`, which places the light at infinity and directs light rays in the direction specified by the light `'position'`. The `lighting` options specify whether the light will be reflected or not. The option, `lighting phong` (other options are `flat`, `gouraud`, and `none`) usually results in the best rendering of reflection, especially from curved surfaces.

Here is a simple example of how to change the edge color, edge transparency, face transparency, and lighting. We use the cylinder plotted previously with the `surf` function and spruce it up with the following commands:

```
% Spruce up the cylinder
light('position',[2 -2 .1],'Style','infinite');  % create a light source
lighting phong;             % define how the light is reflected
set(c,'facealpha',0.8);     % set the face alpha (opacity) to 80%
dark_brown = [.32 .19 .19]; % define a dark_brown color with RGB values
set(c,'edgecolor',dark_brown); % change the edge color to dark brown
set(c,'edgealpha',0.1);     % make the edges almost transparent
set(gca,'visible','off');   % remove the axes and the frame
```

The resulting image is shown in Fig. 6.16(b). Note that the images shown here are in grayscale (to save you money in the cost of this book) and do not really do justice to how stunning they look in color.

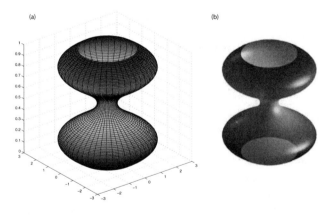

Figure 6.16: Manipulation of 3-D graphics (a) a cylinder generated with `cylinder` and `surf` functions, and (b) the cylinder after changing the face color, edge color, face alpha and edge alpha of the surface patches, and introducing a light source at infinity.

6.5.3   A little more fun with color and lighting

Let us continue just a little longer and explore the variations in face color, edge color, and transparency. This time, let us split the cylinder in two halves (it is really easy to do by manipulating the Z matrix) and render their surfaces differently. We first draw the right half of the cylinder and render it coffee (with milk!) colored. Note the manipulation of the data—columns 15 to 40 have been replaced with NaNs. Here Z is a 101×51 matrix. The rows here represent patches along the z-axis and columns represent the patches along the circumference. You need to figure out which columns you want to retain to plot the part of the cylinder you desire. Of course, you can split it right in the middle and use 3-D rotation to position the half cylinder the way you wish to see it.

```
% Right half of the cylinder
clf                             % clear the figure window
Z1=Z;                           % copy the data for manipulation
Z1(:,15:40)=NaN;                % set part of the data to NaN
c3=surf(X,Y,Z1); axis square;   % plot Z1 to see half of the cylinder
light('position',[2 -2 .1],'Style','infinite');
set(c3,'facecolor',[.99 .92 .80],'facealpha',1);
set(c3,'edgecolor',[.32 .19 .19],'edgealpha',0.1);
lighting phong;
set(gca,'visible', 'off')
view(-38.5,26);
```

The resulting image is shown in Fig. 6.17(a). Note that we have used a `facealpha` of 1 to make the surface opaque here (there is nothing interesting to see through on the other side). We give similar treatment to the left half of the cylinder but color it bright golden and to make it glitter add another light source:

```
% Left half of the cylinder
clf;                            % clear the figure window
ZL=Z;                           % copy the data for manipulation
ZL(:,1:14)=NaN;                 % set part of the data to NaN
ZL(:,41:51)=NaN;
c2=surf(X,Y,ZL); axis square;   % plot Z1 to see half of the cylinder
l1=light('position',[2 -2 .1],'Style','infinite');
l2=light('position',[-2 2 .1],'Style','infinite');
set(c2,'facecolor',[.87 .49 0],'facealpha',1);
set(c2,'edgecolor',[.87 .49 0],'edgealpha',0.1);
lighting phong;
set(gca,'visible','off');
view(-38.5,26);
```

The image of the right half of the cylinder is shown in Fig. 6.17(b). Note the reflection from the surface, clearly indicating light falling from two different directions.

While we are at it, let us plot the bottom half of the cylinder as well, using same color, transparency, and light sources as those for the right half of the cylinder, but use `lighting flat` to alter the reflection characteristics of the patch objects. With flat lighting, each patch has constant reflectance, and hence the individual patches become more visible. The code is follows (for brevity, we have omitted a few lines from this code that you must put in for setting the light source and face color, etc.) and the resulting image is shown in Fig. 6.17(c).

```
% Bottom half of the cylinder (or a 'golden onion')
Z1=Z;
Z1(51:101,:)=NaN;
c2=surf(X,Y,Z1); axis square
% copy code for light, facecolor, etc., here from left half cylinder
lighting flat;
view(-40,20);
```

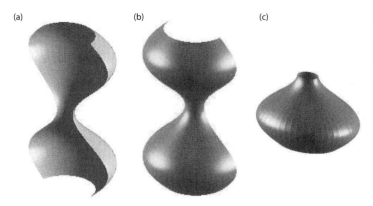

Figure 6.17: Further manipulation of 3-D graphics: the two halves of the cylinder manipulated separately, (a) the right half of the cylinder in coffee color, with no face transparency, (b) left half of the cylinder in golden color, with two independent light sources, and (c) bottom half of the cylinder with lighting set to `flat`.

6.5.4   A word about choosing colors

Colors in MATLAB graphics are specified by their RGB values, a triple, with each number between 0 and 1. A pure red is [1 0 0] and a yellow is [1 1 0]. There are several popular colors predefined—blue, green, red, cyan, magenta, yellow, and black. Thus, it is sufficient to say `plot(x,y,'color','magenta')` rather than `plot(x,y,'color', [1 0 1])`. However, for selecting other colors, it is essential to specify the corresponding RGB values. Now, how do you find these values? In MATLAB's graphics editor, you can open the color palette by clicking on the **color** in the list of items in the **Inspector** window. Select the color you like from **More Colors** and click on the RGB tab to see its RGB values. These values are given as integers that vary from 0 to 255. Note down the numbers and divide them by 255 to get the decimal values between 0 and 1. For example, the golden color is [222 125 0], which must be specified as [.87 .49 0] in the `facecolor` value while plotting the left half of the cylinder.

6.6   Saving and Printing Graphs

For on-line help type:
`help print`

The simplest way to get a hard copy of a graph is to type `print` in the command window after the graph appears in the figure window. The `print` command sends the graph in the current figure window to the default printer in an appropriate form. On PCs (running Windows) and Macs, you could, alternatively, activate the figure window (bring to the front by clicking on it) and then select `print` from the file menu.

The figure can also be saved into a specified file in the PostScript (PS) or Encapsulated PostScript (EPS) format. These formats are available for black and white as well as color printers. The PostScript supported includes both level 1 and level 2 PostScript. The command to save graphics to a file has the form

> `print -d`*devicetype* `-options` *filename*

where *devicetype* for PostScript printers can be one of the following:

devicetype	Description	devicetype	Description
ps	black-and-white PS	eps	black-and-white EPS
psc	color PS	epsc	color EPS
ps2	level 2 black-and-white PS	eps2	level 2 black-and-white EPS
psc2	level 2 color PS	epsc2	level 2 color EPS

For example, the command

> `print -deps sineplot`

saves the current figure in the Encapsulated PostScript file **sineplot.eps**. The **.eps** extension is automatically generated by MATLAB.

The standard optional argument *-options* supported are `append`, `epsi`, `Pprinter`, and `fhandle`. There are several other platform-dependent options. See the on-line help on `print` for more information.

In addition to the PostScript devices, MATLAB supports a number of other printer devices on UNIX and PC systems. There are device options available for HP LaserJet, DeskJet, and PaintJet printers, DEC LN03 printers, Epson printers, and other types of printers. See the on-line help on `print` to check the available devices and options.

Other than printer devices, MATLAB can also generate a graphics file in the following popular formats, among many others (see on-line help on `print`).

`-dill`   saves file in Adobe Illustrator format,
`-djpeg`   saves file as a JPEG image, and
`-dtiff`   saves file as a compressed TIFF image.

The Adobe Illustrator format is quite useful if you want to dress up or modify the figure in a way that is very difficult to do in MATLAB. Of course, you must have access to Adobe Illustrator to be able to open and edit the saved graphs. Figure 6.18 shows an example of a graph generated in MATLAB and then modified in Adobe Illustrator.

6.6.1   Saving graphs to reusable files

It is also possible to save a graphics file as a list of commands, regenerate the graphics later, and modify it. To save the graphics in the currently active window, type

hgsave filename.fig

Later, you can open the file with the command **open** <i>filename</i>.fig to get the plot back into the graphics window. The **hgsave** command saves the plot with its handles and the associated data. You can also save the plot as a figure file (<i>filename</i>.fig) by selecting Save or Save As... from the File menu in the graphics window. Alternatively, you could use the command saveas(gcf, '<i>filename</i>','fig') on the command line.

For on-line help type:
`help hgsave`
`help saveas`

There is yet another way of saving a graph in a file that can be used in MATLAB to recreate the graph. You can select **Create M-code** from the File menu of the graphics window. This action will create a function file, an M-file, that will require input data to recreate the graph. Thus, the generated M-code contains all commands necessary to create the graph but does not contain the data that was used for the graph. This particular way of saving a graph may be useful when you make a lot of changes to your graph using plot editor and you would like to save the final settings for creating similar plots with possibly different data.

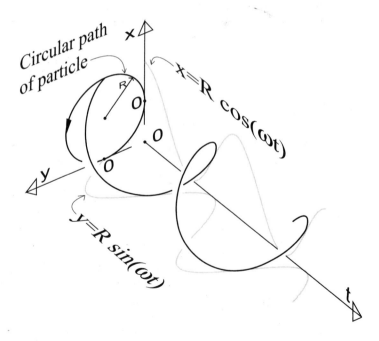

Figure 6.18: Example of a figure generated in MATLAB, saved in the Illustrator format and then modified in Adobe Illustrator. The rotation and shearing of texts was done in Illustrator (courtesy of A. Ruina).

6.7 Animation

We all know the visual impact of animation. If you have a lot of data representing a function or a system at several time sequences, you may wish to take advantage of MATLAB's capability to animate your data.

There are three types of facilities for animation in MATLAB.

1. **Comet plot:** This is the simplest and the most restricted facility to display a 2-D or 3-D line graph as an animated plot. The command `comet(x,y)` plots the data in vectors x and y with a comet moving through the data points. The trail of the comet traces a line connecting the data points. So, rather than having the entire plot appear on the screen at once, you can see the graph "being plotted." This facility may be useful in visualizing trajectories in a phase plane. For an example, see the built-in demo on the Lorenz attractor.

2. **Movies:** If you have a sequence of plots that you would like to animate, use the built-in `movie` facility. The basic idea is to store each figure as a frame of the movie, with each frame stored as a column vector of a big matrix, say M, and then to play the frames on the screen with the command `movie(M)`. A frame is stored in a column vector using the command `getframe`. For efficient storage, you should first initialize the matrix M. The built-in command `moviein` is provided precisely for this initialization, although you can do it yourself too. An example script file to make a movie might look like this:

```
%--- skeleton of a script file to generate and play a movie  ---
%
nframes = 36;              % number of frames in the the movie
Frames = moviein(nframes); % initialize the matrix 'Frames'
for i = 1:nframes
     :                     % you may have calculations here to
     :                     %- generate data
     :
    x = ....;
    y = ....;
    plot(x,y)              % you may use any plotting function
    Frames(:,i) = getframe; % store the current figure as a frame
end
movie(Frames,5)           % play the movie Frames 5 times
```

You can also specify the speed (frames/second) at which you want to play the movie (the actual speed will eventually depend on your CPU) by typing `movie(Frames, m, fps)`, which plays the movie, stored in *Frames*, m times at the rate of *fps* frames per second.

3. **Handle Graphics:** Another way, and perhaps the most versatile way, of creating animation is to use the Handle Graphics facilities. The basic idea here is to plot an object on the screen, get its handle, use the handle to change the desired properties of the object (most likely its *xdata* and *ydata* values), and replot the object over a selected sequence of times. There are

two important things to know to be able to create animation using Handle Graphics:

- The command `drawnow`, which flushes the graphics output to the screen without waiting for the control to return to MATLAB. The on-line help on `drawnow` explains how it works.

- The object property `erasemode`, which can be set to `normal`, `background`, `none`, or `xor` to control the appearance of the object when the graphics screen is redrawn. For example, if a script file containing the following lines is executed

```
h1 = plot(x1,y1,'erasemode','none');
h2 = plot(x2,y2,'erasemode','xor');
:
newx1 = ...;
newy1 = ...;
newx2 = ...;
newy2 = ...;
:
set(h1,'xdata',newx1,'ydata',newy1);
set(h2,'xdata',newx2,'ydata',newy2);
:
```

then the first `set` command draws the first object with the new x-data and y-data, but the same object drawn before remains on the screen, while the second `set` command redraws the second object with new data and also erases the object drawn with the previous data x_2 and y_2. Thus, it is possible to keep some objects fixed on the screen while some other objects change with each pass of a control flow.

Now let us look at some examples that illustrate the use of Handle Graphics in animation.

Example 1: A bead goes around a circular path

The basic idea is to first calculate various positions of the bead along the circular path, draw the bead as a point at the initial position and create its handle, and then use the handle to set the x- and y-coordinates of the bead to new values inside a loop that cycles through all positions (see Fig. 6.19). The `erasemode` property of the bead is set to `xor` (Exclusive Or) so that the old bead is erased from the screen when the new bead is drawn. Try the following script file.

```
% Script file for animating the circular motion of a bead
% --------------------------------------------
clf                             % clear any previous figure
theta = linspace(0,2*pi,1000);    % create a vector theta
x = cos(theta);                 % generate x and y-coordinates
y = sin(theta);                 %- of the bead along the path
hbead = line(x(1),y(1),'marker','o',...
   'markersize',8,'erase','xor'); % draw the bead at the initial
                                %- position and assign a handle
axis([-1 1 -1 1]); axis('square');
for k = 2:length(theta)         % cycle through all positions
   set(hbead,'xdata',x(k),'ydata',y(k));
                                % draw the bead at the new position
   drawnow
end
```

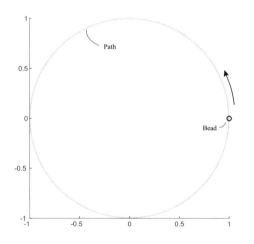

Figure 6.19: A bead goes on a circular path.

Example 2: The bead going around a circular path leaves its trail

In Example 1, the bead goes on the circular path, but it does not clearly seem that it traverses a circle. To make it clear, we can make the bead leave a trail as it moves. For this purpose, we basically draw the bead twice, once as a bead (with bigger marker size) and once as a point at each location (see Fig. 6.20). But we set the `erasemode` property of the point to `none` so that the point (the previous position of the bead) remains on the screen as the bead moves and thus creates a trail of the bead.

```
% Script file for animating the circular motion of a bead. As the
% bead moves, it leaves a trail behind it.
% ---------------------------------------------
clf
theta = linspace(0,2*pi,1000);
x = cos(theta); y=sin(theta);
hbead = line(x(1),y(1),'marker','o','markersize',8,'erase','xor');
htrail = line(x(1),y(1),'marker','.','color','r','erase','none');
axis([-1 1 -1 1]);
axis('square');
for k = 2:length(theta)
    set(hbead,'xdata',x(k),'ydata',y(k));
    set(htrail,'xdata',x(k),'ydata',y(k));
    drawnow
end
```

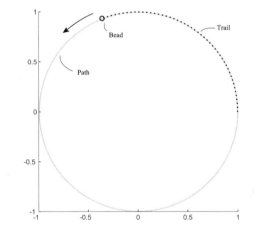

Figure 6.20: A bead goes on a circular path and leaves a trail behind it.

Example 3: A bar pendulum swings in 2-D

Here is a slightly more complicated example. It involves animation of the motion of a bar pendulum governed by the ODE $\ddot{\theta} + \sin\theta = 0$. Now that you are comfortable with defining graphics objects and using their handles to change their position, etc., the added complication of solving a differential equation should not be too hard.

```
%----- script file to animate a bar pendulum --------------

clf                        % clear figure and stuff
data = [0 0; -1.5 0];      % coordinates of endpoints of the bar
phi = 0;                   % initial orientation
R = [cos(phi) -sin(phi);   sin(phi) cos(phi)];
                           % rotation matrix
data = R*data;
axis([-2 2 -2 2])          % set axis limits
axis('equal')

%-----define the objects called bar, hinge, and path.
bar = line('xdata',data(1,:),'ydata',data(2,:),...
          'linewidth',3,'erase','xor') ;
hinge = line('xdata',0,'ydata',0,'marker','o','markersize',[10]);
path = line('xdata',[],'ydata', [],'marker','.','erasemode','none');

theta = pi-pi/1000;        % initial angle
thetadot = 0;              % initial angular speed
t = 0;                     % initial time
dt = .2;                   % time step
tfinal = 50;               % final time

%------Euler's method for numerical integration
while(t<tfinal);
    t=t+dt;
    theta=theta + thetadot*dt;
    thetadot=thetadot -sin(theta)*dt;
    R=[cos(theta)  -sin(theta); sin(theta) cos(theta)];
    datanew= R*data;

    %---- change the property values of the objects: path and bar.
    set(path,'xdata', datanew(1,1), 'ydata', datanew(2,1) );
    set(bar,'xdata',datanew(1,:),'ydata',datanew(2,:) );
    drawnow;
end
```

Example 4: The bar pendulum swings, and other data are displayed

Now here is the challenge. If you can understand the following script file, you are in good shape! You are ready to do almost any animation. The following example divides the graphics screen in four parts, shows the motion of the pendulum in one part, shows the position of the tip in the second part, plots the angular displacement θ in the third part, and plots the angular speed $\dot{\theta}$ in the fourth part (see Fig. 6.21). There are four animations occurring simultaneously. Try it! There is an intentional bug in one of the four animations. If you start the pendulum from the vertical upright position with an initial angular speed (you will need to change `thetadot=0` statement inside the program for this), then you should see the bug. Go ahead, find the bug and fix it.

```
%----- script file to animate a bar pendulum and the data --------

% get basic data for animation
% ask the user for initial position
disp('Please specify the initial angle from the')
disp('vertical upright position.')
disp(' ')
offset = input('Enter the initial angle now: ');
                              % ask the user for time of simulation
tfinal = input('Please enter the duration of simulation: ');
disp('I am working....')

theta = pi-offset;          % initial angle
thetadot = 0;               % initial angular speed
dt = .2;  t=0;  tf=tfinal;  % time step, initial and final time

disp('Watch the graphics screen')
clf                         % clear figure and stuff
h1 = axes('position',[0.55 .1 .4 .3]);
axis([0 tf -4 4]);          % set axis limits
xlabel('time'),ylabel('displacement')
Displ = line('xdata',[],'ydata', [],'marker','.','erasemode','none');

h2 = axes('position',[0.55 .55 .4 .3]);
axis([0 tf -4 4]);          % set axis limits
xlabel('time'),ylabel('velocity')
Vel = line('xdata',[],'ydata', [],'marker','.','erasemode','none');

h3 = axes('position',[.1 .1 .4 .4]);
axis([-pi pi -4 4])         % set axis limits
axis('square')
xlabel('displacement'),ylabel('velocity')
Phase = line('xdata',[],'ydata', [],'marker','.','erasemode','none');

h4 = axes('position',[.1 .55 .4 .4]);
axis([-2 2 -2 2])           % set axis limits
axis('square')
```

```
data = [0 0; -1.8 0];           % coordinates of endpoints of the bar
phi = 0;                        % initial orientation
R = [cos(phi) -sin(phi); +sin(phi) cos(phi)];   % rotation matrix
data=R*data;
%-----define the objects called bar, hinge, and path.
bar = line('xdata',data(1,:),'ydata',data(2,:),...
           'linewidth',3,'erase','xor');
hinge = line('xdata',0,'ydata',0,'marker','o','markersize',[10]);
path = line('xdata',[],'ydata', [],'marker','.','erasemode','none');
%------Euler's method for numerical integration
while(t<tfinal);
    t = t+dt;
    theta = theta + thetadot*dt;
    thetadot = thetadot -sin(theta)*dt;
    R = [cos(theta) (-sin(theta)); sin(theta) cos(theta)];
    datanew = R*data;
    %---- change the property values of the objects: path and bar.
    set(path,'xdata', datanew(1,1), 'ydata', datanew(2,1) );
    set(bar,'xdata',datanew(1,:),'ydata',datanew(2,:) );
    set(Phase,'xdata', theta, 'ydata', thetadot);
    set(Displ,'xdata', t, 'ydata', theta );
    set(Vel,'xdata', t, 'ydata', thetadot );
    drawnow;
end
```

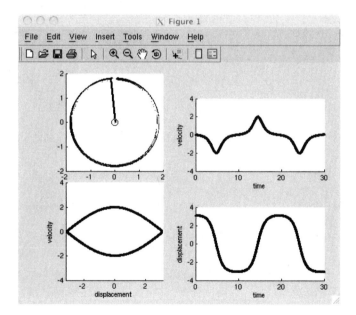

Figure 6.21: Animation of motion of a bar pendulum along with animation of position and velocity data.

7. *Errors*

Errors are an integral part of life whether you interact with computers or not. The only difference is, when you interact with computers, your errors are pointed out immediately—often bluntly and without much advice. Interaction with MATLAB is no exception. Yes, to err is human, but to forgive is definitely not MATLABine. So, the earlier you get used to the blunt manners of your friend and his terse comments, the better for you. As this friend does not offer much advice most of the time, we give you some hints here based on our own experience in dealing with your friend. Before we begin, we warn you that this friend has a tendency to become very irritating if you work under too much time pressure or don't have enough sleep. In particular, if you are not relaxed enough to distinguish between (and [, ; and :, or a and A, you and your friend are going to have long sessions staring at each other.

Here are the most common error messages that you are likely to get while working in MATLAB. All messages below are shown following a typical command. Following the actual message are explanations and tips.

1.
```
>> D = zeros(3); d = [1 2];

>> D(2:3,:) = sin(d)

???  Subscripted assignment dimension mismatch.
```

This is a typical problem in matrix assignments where the dimensions of the matrices on the two sides of the equal sign do not match. Use the `size` command to check the dimensions on both sides and make sure they agree. For example, for the preceding command to execute properly, `size(D(2:3,:))` and `size(sin(d))` or `size(d)` must give the same dimensions.

2.
```
>> (x,y) = circlefn(5);

??? (x,y)=circlefn(5);
        |
Error: Expression or statement is incorrect--possibly
        unbalanced (, {, or [.
```

Here MATLAB is confused with the list of variables because a list within *parentheses* represents matrix indices. When the variables represent output of a function or a list of vectors, they must be enclosed within *square brackets*. The correct command here is [x,y]=circlefn(5) (circlefn is a user-written function). When parentheses and brackets are mixed up, the same error message is displayed:

```
>> (x,y] = circlefn(5);

??? (x,y]=circlefn(5);
        |
Error: Expression or statement is incorrect--possibly ...
```

3.
```
>> x = 1:10;

>> v = [0 3 6];

>> x(v)

??? Subscript indices must either be real positive integers
or logicals.
```

The first element of the index vector v is zero. Thus, we are trying to get the zeroth element of x. But zero is not a valid index for any matrix or vector in MATLAB except when it is a logical zero (i.e., zeros produced by logical operations). The same problem arises when a negative number appears as an index. Also, of course, an error occurs when the specified index exceeds the corresponding dimension of the variable:

```
>> x(12)

??? Attempted to access x(12); index out of bounds because numel(x)=10.
```

The examples given here for index-dimension mismatch are almost trivial. Most of the times, these problems arise when matrix indices are created, incremented, and manipulated inside loops.

4.
```
>> x=1:10; y=10:-2:-8;

>> x*y

??? Error using ==> mtimes

Inner matrix dimensions must agree.
```

In matrix multiplication x*y, the number of rows of x must equal the number of columns of y. Here x and y are both row vectors of size 1×10 and therefore cannot multiply. However, x*y' and x'*y will both execute without error, producing inner and outer products, respectively.

Several other operations involving matrices of improper dimensions produce similar errors. A general rule is to write the expression on paper and think if the operation makes mathematical sense. If not, then MATLAB is likely to give you error. For example, A^2 makes sense for a matrix A only if the matrix is square, and A^x for a vector x and matrix A does not make any sense. The exceptions to this rule are the two division operators—/ and \. Although y/x, for the two vectors defined earlier, may not make any sense mathematically, MATLAB gives an answer:

```
>> y/x

ans =
    -0.2857
```

This is because this division is not a mere division in MATLAB, but it also gives solutions to matrix equations. For rectangular matrices, it gives solutions in the least squares sense. See on-line help on **slash** for more details.

A common source of error is to use the matrix operators where you want array operators. For example, for the vectors x and y, y.^x gives element-by-element exponentiation, but y^x produces an error:

```
>> y^x

??? Error using ==> mpower

At least one operand must be scalar.
```

5.
```
>> [x,y] = circlefn;
??? Input argument "r" is undefined.

Error in ==> CIRCLEFN at 9
x = r*cos(theta);                    % generate x-coordinates
```

A function file has been executed without giving proper input. This is one of the very few error messages that provides enough information (the function name, the name of the directory where the function file is located, and the line number where the error occurred).

6.
```
>> [t,x] = Circle(5);

??? Attempt to execute SCRIPT circle as a function.
```

Here `Circle` is a script file. Input-output lists cannot be specified with script files. But here is a slightly more interesting case that produces the same error. The error occurs in trying to execute the following function file:

```
Function [x,y] = circlefn(r);
% CIRCLEFN - Function to draw a circle of radius r.
theta = linspace(0,2*pi,100);        % create vector theta
x = r*cos(theta);
y = r*cos(theta);                     % generate coordinates
plot(x,y);                            % plot the circle
```

Here is the error:

```
>> [x,y]=circlefn(5)

??? Attempt to execute SCRIPT circlefn as a function.
```

You scream, "Hey, it's not a script!" True, but it is not a function either. For it to qualify as a function, the f in `function`, in the definition line, must also be in lowercase, which is not the case here. So MATLAB gets confused. Yes, the error message could have been better, but you are probably going to say this in quite a few cases.

```
>> CIRCLEFN[5];

??? circlefn[5]
           |
Error: Unbalanced or unexpected parenthesis or bracket.
```

Here parentheses are required in the place of the square brackets. The error locator bar is at the right place and should help. This error message is good and helpful. We have indeed misused brackets!

```
>> EIG(D)

??? Undefined function or method 'EIG' for input arguments of type ...
```

Here `EIG` is a built-in function and its name must be typed in lowercase: `eig`. MATLAB does not recognize `EIG`, hence the error. But the error message provides no clue about the existence of `eig`. In previous versions of MATLAB, this error was better diagnosed and told the user that it was a capitalized internal function.

7.
```
>> x = b+2.33

??? Undefined function or variable b.
```

The variable b has not been defined. This message is right on target. But when the same message comes for a function or script that you have written, you may scratch your head. In such cases, the function or the script that you are trying to execute is most probably in a different directory than the current directory. Use `what`, `dir`, or `ls` to show the list of files in the current directory. If the file is not listed there then MATLAB cannot access it. You may have to locate the directory of the file with the command `which` *filename* and then change the working directory with the `cd` command to the desired directory.

8.
```
>> global a, b

??? Undefined function or variable b.
```

You think that you are merely declaring a and b to be global variables. But what is that comma doing there (after `a`)? MATLAB treats the comma to be a command or statement separator. Thus, it did accept the declaration `global a` as a valid command; now it is looking for a value of b (because typing a variable by itself directs MATLAB to return its value). So, do not use commas to separate variables in `global`, `save`, or `load` commands. A really tricky situation arises when you make a similar mistake in a `for` loop statement: `for i=1, n`. This statement will execute without any error message, but the loop will be executed for only $i = 1$ and MATLAB will happily display the value of n on the screen.

9.
```
>> d = 0:10;
>> d1 = linspace(0,2*pi,10);
>> plot(d,d1)

??? Error using ==> plot
Vectors must be the same lengths.
```

The input vectors in `plot` command must be pairwise compatible. For a detailed discussion of this, see the description of the plot command in Chapter 6.

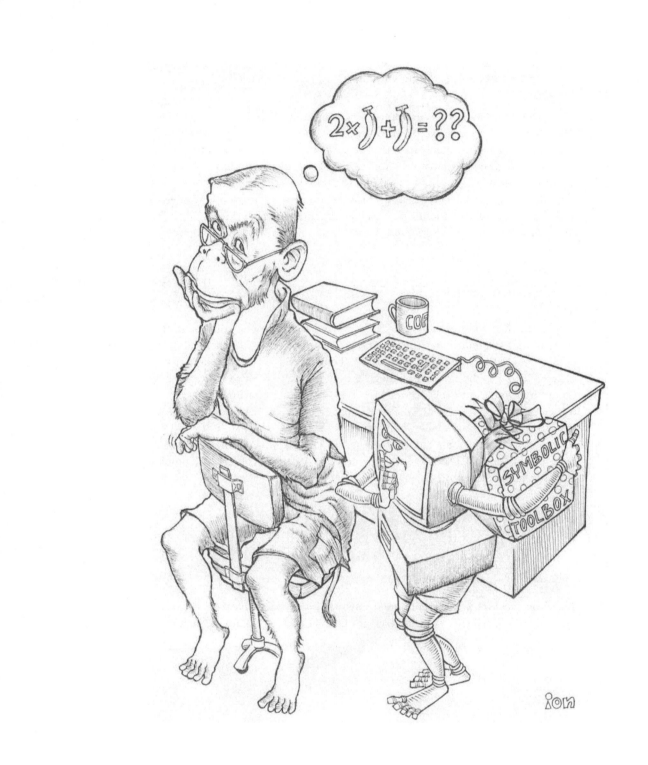

8. Computer Algebra and The Symbolic Math Toolbox

Computer algebra refers to mostly non-numerical mathematical computation on computers where symbols are used rather than numbers. In mathematical modeling and analysis, we mostly work with symbolic variables, e.g., x, y, α, β; write equations using these variables; and try to obtain our final answers in terms of these variables. Computer algebra is meant for such calculations. It uses very different methods internally for such computations than what numerical computation techniques use. There are dedicated packages, such as Mathematica, Maple, Macsyma (almost extinct now), and MuPAD, that do computer algebra. Because MATLAB is a package for numerical computation, it cannot do computer algebra without a separate computational engine to drive it. Symbolic Math Toolbox does precisely that—it provides a gateway to a computer algebra package. Until 2007, the engine inside Symbolic Math Toolbox was Maple; now it is MuPAD.

8.1 The Symbolic Math Toolbox

The Symbolic Math Toolbox allows MATLAB to respond to MuPAD commands. MATLAB is a program that does mostly arithmetic and MuPAD is another program that does mostly algebra and calculus. MATLAB mostly gives output that is a number or an array of numbers. MuPAD is designed to give output in symbolic form. Matlab can tell you that $\sqrt{2}$ is about 1.41415 and MuPAD can tell you that the solutions to $x^2 = c$ are $x = \pm\sqrt{c}$ and leave it at that with the letter c. MATLAB comes from the numeric calculation tradition of Fortran, Basic, and C. MuPAD comes from the artificial intelligence tradition of (first and some say best)

Macsyma, later imitated by Mathematica and others. In the recent past, computer algebra packages have added the ability to give numeric solutions and make nice plots. Even though these programs run slower, are harder to learn, and are harder to use than MATLAB for these purposes, Mathworks has felt the competition and responded by providing this toolbox.

To master the Symbolic Math Toolbox, you have to master MuPAD, which is at least as much work as learning MATLAB. You also have to learn how to run MuPAD from inside MATLAB and keep your head straight at the same time. Basically, you type a command that makes sense to either plain MATLAB or MuPAD, and if plain MATLAB can't make sense of what you are requesting, it checks to see if you have typed a legitimate MuPAD command (i.e., MATLAB's version of the MuPAD command).

8.1.1   Should you buy it?

Although some people swear by them, symbolic calculations seem to be less useful than numeric computation to many people for two reasons. First, many problems are too hard to do symbolically. An infinitely fast computer doing symbolic computation will never tell you a formula for the integral of $\sin(\log(x)^2)$, but MATLAB will tell you the area under the curve as accurately as you would like (see Section 5.4, page 152). Second, the symbolic formulas one obtains are often unwieldy. How does, say, a 50-term symbolic formula give one more insight than a numeric calculation?

With the Symbolic Math Toolbox, MATLAB can do essentially all that Mathematica, Maple, or MuPAD can do. Because the Symbolic Math Toolbox *comes* with the student edition of MATLAB, you might as well use MuPAD there for free. If you don't have the Symbolic Math Toolbox and you don't know of any symbolic calculations that you need to do, wait. Most engineers can do fine without it.

If you want to do a lot of back and forth between symbolic and numeric computation, then MATLAB's Symbolic Math Toolbox is just the thing for you.

8.1.2   Two useful tools in the Symbolic Math Toolbox

Before getting more involved, you should note two cute and useful tools that come with the Symbolic Math Toolbox.

1. **A quick way to make plots: ezplot.** Here is the simplest way to make simple plots. This function (along with its cousins) is also available in basic MATLAB (see Section 3.8 on 92 for examples).

   - The command `ezplot sin(t)`, or more elaborately, `ezplot('sin(t)')`, makes a plot of the sin function on the default interval $(-2\pi, 2\pi)$.
   - The command `ezplot('t^2/exp(t)', [-1 5])` plots the function t^2/e^t over the interval $(-1, 5)$.

If you find that you want more control over your plots than `ezplot` gives you, learn more of MATLAB's plotting features in Lesson 3 of Chapter 2 and in Chapter 6.

2. **A fun tool:** `funtool`. Type `funtool` and you will be operating a two-screen plotting calculator that does symbolic calculations. Use this when you need to do some quick checking on a function, its derivative, integral, inverse, etc. The help key explains what the other keys do. Some correspond to basic MuPAD commands that you can use at the MATLAB command line as well.

8.2   Numeric Versus Symbolic Computation

What is the difference between numeric (plain MATLAB) and symbolic (the Symbolic Math Toolbox) computation? Let us say you wanted to know the derivative of a function, $t^2 \sin(3t^{1/4})$.

The definition of derivative gives a crude way to *numerically* calculate the derivative of a function as $f' \approx \{f(x+h) - f(x)\}/h$ if h is small. The MATLAB commands that follow calculate this approximation of the derivative at a set of points spaced $\Delta t = h$ apart.

```
h =.1;                      % delta t for the  difference
t = -pi : h : pi;           % the region of interest for t
f = t.^2.*sin(3*t.^(1/4));  % the function at all the t values
fprime = diff(f)/h;         % numerical approx of derivative
                % Note: diff(f) has 1 less element than f
plot(t(1:end-1), fprime)    % plot of the derivative
```

The derivative of the function is represented by the two lists of numbers t and *fprime*. The derivative at $t(7)$ is *fprime*(7).

Compare this with the Symbolic Math Toolbox calculation of the derivative of the same function. Here is the command and response.

```
>> symb_deriv = diff('t^2 * sin(3*t^(1/4))')

symb_deriv =

2*t*sin(3*t^(1/4))+3/4*t^(5/4)*cos(3*t^(1/4))
```

The Symbolic Math Toolbox gives you a formula for the derivative. If you are patient you can verify by hand that indeed

$$\frac{d}{dt}\left(t^2 \sin(3\,t^{1/4})\right) = 2\,t\sin(3\sqrt[4]{t}) + (3/4)\,t^{5/4}\cos(3\sqrt[4]{t}).$$

A plot of the previous function gives about the same curve as the previous numeric calculation. If you type `int(symb_deriv)`, you will get back `t^2*sin(3*t^(1/4))` as expected[1] by the fundamental theorem of calculus. MuPAD is pretty good at basic calculus.

Notice that `diff` is two different commands. One in MATLAB (dealing with differences between consecutive items in a list) and one in MuPAD (symbolically calculating the derivative). Whether MATLAB or MuPAD responds when you use `diff` depends on whether you have typed something in the correct syntax for one or the other program.

8.2.1   Variable precision arithmetic

All computer algebra packages, including MuPAD, can do variable precision arithmetic: that is, calculate numbers to, theoretically, any arithmetic precision. This is in contrast to the numbers calculated by MATLAB that are limited to double precision. The increase in precision, however, comes at a cost of significant increase in computational time. The command to increase the precision is remarkably simple. All you have to do is issue the command

For on-line help type:
`help vpa`

$$\text{vpa}(symbolic\_expression,\ no\_of\_digits)$$

to evaluate the *symbolic_expression* up to the desired precision. Thus,

`vpa(sin(pi/4))`   evaluates to 0.70710678118654752440084436210485
                                        (default precision of 32 digits), and
`vpa(sin(pi/4), 40)` evaluates to 0.7071067811865475244008443621048490392848
                                        (desired precision of 40 digits),

whereas the plain MATLAB evaluates `sin(pi/4)` to be 0.707106781186547 using its default double-precision numerical accuracy.

Thus, you have at least two different options in numerical evaluation of quantities of interest: (1) as a double-precision floating-point number (MATLAB default) using `double`, or (2) as a variable precision number using `vpa`. You need to be careful when you use functions such as `sqrt` on numbers, which by default result in a double-precision floating-point number. You need to pass such input to `vpa` as a symbolic string for correct evaluation: `vpa('sqrt(5)/pi')`. [A teaser: Try evaluating `double('sqrt(5)/pi')`. What does the answer mean? Hint: See on-line help on `double`.]

8.3   Getting help with the Symbolic Math Toolbox

There are many ways of getting on-line help on MuPAD. These few pages are just the tip of an iceberg. To go further without going to the library or bookstore, you can get help from MATLAB's usual labyrinth of help options.

- If you know the name of the command you want help with, for instance, `solve`, you can see helpful explanations in any of three ways:

---

[1]Well, not quite. You will have to use `simplify` after the integration to get back the original expression. Alternatively, you could use `simplify(int(symb_deriv))`.

1. Type `help solve` at the command line. Luckily, `solve` is a MuPAD command and is not also a plain MATLAB command. Beware, some commands have meaning to both plain MATLAB and the Symbolic Math Toolbox, such as the command `diff` used earlier. If you type `help diff`, you will see a description of the plain old MATLAB command `diff` with this helpful clue at the end:

   ```
   Overloaded methods
       help sym/diff.m
       help char/diff.m
   ```

   `Overloaded` means the command `diff` has meaning outside of plain MATLAB. And `help sym/diff.m` means that if you type `help sym/diff` you will get help with the Symbolic Math Toolbox (i.e., MuPAD) command, also called `diff`.

2. Type `help sym/solve` at the command line to get the same help file as `help solve`. Type `help sym/diff` to get help on the *symbolic* command `diff`.

3. Type `helpdesk` at the MATLAB prompt. Then click on **Symbolic Math Toolbox** in the **Help Navigator** pane of the **Help** window. Select the help resource that interests you (**Getting Started** is a good first choice).

- To see an organized list of MuPAD commands with a very short description you can do either of two things:

  1. Type `help symbolic` on the command line. One line in this list is, for example,
     "`solve - Symbolic solution of algebraic equations.`"

  2. Click on the ⊘ button at the top of the command window. The **Help** window opens up. Select **Symbolic Math Toolbox** in the **Help Navigator** pane and then click on **By Category** under **Functions** in the main **Help** window. You will be presented with the same list of functions but in the help window.

- "Live" demonstrations of a few groups of commands are available. These demos take about a minute if you gloss your eyes and repeatedly hit the space bar. They take 10–30 minutes if you try to follow them carefully. The demos are `symintro` (introduction), `symcalcdemo` (calculus), `symlindemo` (linear algebra), `symvpademo` (variable precision arithmetic), `symrotdemo` (rotations in the plane), and `symeqndemo` (solution of equations). There are two ways to get to these demos:

  1. Type, say, `symcalcdemo` on the command line and then do as told.

  2. Type `demos` at the command line. Then click on **Symbolic Math**, then click on, say, **Introduction**, and then click on the topic of your choice.

- The 450+-pages-long *Symbolic Math Toolbox User's Guide* is on your computer if someone installed it. At the command line, type helpdesk. In the web browser that pops up, select Symbolic Math Toolbox, click on the User's Guide under the Documentation Set and navigate yourself through various options. You can also download the PDF file of the User's Guide by selecting it under the Printable (PDF) Documentation On the Web category.

8.4 Using the Symbolic Math Toolbox

Let's see some other things the Symbolic Math Toolbox can do, besides diff, int, ezplot, and funtool.

8.4.1 Basic manipulations

Expand a polynomial, make it look nice, solve it, and check the solution. Try typing the following lines (one at a time or in an M-file) and keep track of MATLAB's response.

```
syms x  a                 % tell matlab that x and a are symbols
f    = (x-1) * (x-a) * (x + pi) * (x+2) *(x+3)   % define f
g    = expand(f)          % rewrite f, multiplying everything out
h    = collect(g)         % rewrite again by collecting terms
soln = solve(h,x)         % find all the solutions,
check = subs(f,x,soln(5)) % check, say, the fifth solution
```

Comments:

- The syms declaration can sometimes be skipped. MATLAB can sometimes figure out by context that you want a letter to be a symbol. Two ways to do this are with the sym command and with single quotes ('). But it is safest to be explicit and use syms.

- Because x is a symbol, f is automatically treated as a symbolic expression.

- solve is a powerful command that tries all kinds of things to find a solution. Here, solve manages to find all five roots of a fifth-order polynomial (something that cannot always be done, by the way).

- subs is an often-used command if you do math on-line. Here every occurrence of x in the expression f is replaced with the fifth supposed solution. The result of this line of calculation is, predictably, 0.

8.4.2 Talking to itself

Get plain MATLAB to understand the output of the Symbolic Math Toolbox. One confusion is that the output of symbolic commands are symbolic expressions. These often *look exactly* like regular MATLAB expressions but MATLAB doesn't see them that way. There are a few tricks to getting symbolic expressions into a form that you can easily use with plain MATLAB. The key commands are double (turns a symbolic array of numbers into plain old numbers), `eval` (takes text that looks like a MATLAB command and executes the command), and `vectorize` (rewrites a formula so that it can be applied to a whole array of numbers). Try the following commands.

```
syms x t y a                    % Set x,t,y, and a to be symbols
f   = x + sin(x)                % define f(x)
q   = 3*t^2 -7^t                % build up a horrible formula
g   = subs(f,x,q )              % Substitute q(t) in for x
h   = subs(g,t, 'exp(y/a)')     % Substitute exp(y/a) in for t
pretty(h)                       % Print it in a readable form

result = subs(h,{y,a},{7,9})
                                % Evaluate it with y=7 and a=9
a_number_please = double(result)
                                % Get a number you can work with!

y=0:.1:1;   a = pi;             % Try an array of values -
                                % for y and a
y=sym(y); a=sym(a);             % treat the values of -
                                % y and a as symbols

hvec   = vectorize(h)           % Write a formula that works -
                                % with arrays
result = eval(hvec)'            % Evaluate that vectorized -
                                % formula (mess!)

result_numeric =  double(result)    % MuPAD's exact --> numbers
plot (double(y), result_numeric)    % Graph the horrible formula
```

Comments:

- Using substitution, it is easy to build up big messy formulas such as the formula stored in h.

- The command `pretty` sometimes can help you see through a messy formula. Try also `simplify` to reduce formulas using common trig identities, collect common terms, and so on.

- We have used a fancier syntax for `subs` here by substituting two things at once.

- MuPAD writes formulas in a reasonable way for functions of one variable. MATLAB is set up for matrices. To get MATLAB to plug in a formula for an array of numbers (and not perform matrix operations), you have to use array operators (such as `y.^2` to square all the elements of y). The `vectorize` command takes a formula that is good for scalars and puts dots in the right places.

  For on-line help type:
  help vectorize

- When to use (or not) the `double` or `eval` commands is perfectly confusing. Trial and error will be an inevitable part of your work.

8.4.3   Generating MATLAB code for an inline or anonymous function

Sometimes it is convenient to have a new "function" to work with, but you don't want to write a whole M-file for the purpose. You would like to be able to type `myfun(7)` and have a big formula evaluated. In particular, you might like this formula to be one you cooked up with the Symbolic Math Toolbox. So, you need to create either an anonymous function or an inline function (see Section 3.5 on page 83) from the symbolic expression.

Say you want to know how one of the roots of a cubic polynomial depends on one of the coefficients. Here is one approach.

```
syms x a
f = x^3 + a* x^2 + 3*x +5    % A cubic  with parameter a.
roots = solve(f,x)           % Find the three roots (a mess!).
root1 = roots(1)             % Pick out the first root (a mess!).

myfun = inline(char(root1)) % Make an inline function.
myfun(7)                     % Find the root when a=7.
subs(f,{x,a},{ans,7})        % Check the root at a=7.
```

For on-line help type:
`help solve`

For on-line help type:
`help inline`

Inline function creation with the `inline` command has certain limitations. It expects strictly a character string as the input (see comments at the end). Therefore, converting `roots` into an inline function directly is hard (`roots` is a symbolic array). However, creating an anonymous function using the more powerful utility function `matlabFunction` is much easier. Try the following commands in continuation with the previous commands.

For on-line help type:
`help matlabFunction`

```
my_anony_fun = matlabFunction(root1) % Make an anonymous function for root1.
my_anony_fun(7)              % Find the root when a=7.
subs(f,{x,a},{ans,7})        % Check the root at a=7.

my_anony_fun = matlabFunction(roots)
                             % Make an anonymous function for all roots.
my_anony_fun(7)              % Find the roots when a=7.
subs(f,{x,a},{ans(2),7})     % Check out the 2nd root at a=7.
```

Comments:

- `root1` is the symbolic expression for the first root of the cubic polynomial in terms of the parameter a.

- The `inline` function wants a character (string) expression, not a symbolic expression (even though they look the same when typed out), so you have to convert the expression using the `char` function.

- If you want to plug in a list of values for a all at one time, you can change the last two lines as follows:

```
myfun = inline( char(vectorize(root1)) )
myfun(4:.2:8)'           % a, from 4 to 8.
```

8.4.4 Generating M-files from symbolic expressions

When an expression obtained from some symbolic calculation is very long or is of repeated use to you across multiple MATLAB sessions, you will be better served by converting that expression into a MATLAB function, an M-file, rather than an inline or anonymous function. Of course, this conversion is useful only if you are interested in numerically evaluating that expression. Making the Symbolic Math Toolbox generate an M-file for you for a given symbolic expression is fairly easy—use the `matlabFunction` command with additional input specifying a file name:

$$f = \texttt{matlabFunction}(\textit{expression}, \texttt{'file'}, \textit{'file\_name'}\,)$$

Here, `f` becomes the handle of the generated function. The input list depends on the *expression* or explicit specification with additional input parameters to the function `matlabFunction` (see on-line help). The default output variable name is RESULT. Once the file is created, you can edit it like any other M-file.

Here is one example of using `matlabFunction` to generate M-files.

```
syms x a
A = [x a; a*x x^2]        % A is a 2 by 2 matrix
[V, D] = eig(A);          % compute eigen pairs of A.
lambda = diag(D);         % eigenvalues are on the diagonal of D

h_eval = matlabFunction(lambda,'file','A_eigenvalues')
                          % lambda is coded in A_eigenvalues.m.
help A_eigenvalues        % try online help on this function.
A_eigenvalues(1,2)        % find eigenvalues for x=1, a=2.

h_evec = matlabFunction(V,'file','A_eigenvectors')
                          % V is coded in A_eigenvectors.m.
v = A_eigenvectors(1,2)   % find eigenvectors for x=1, a=2.
```

8.5 Using MuPAD Notebook

MuPAD is a powerful package for computer algebra. Apart from hundreds of functions for doing algebra and calculus, it has a very impressive graphics capability. Only a small subset of that capability is accessible through the Symbolic Math Toolbox interface. However, MATLAB lets you access the full might of MuPAD, bypassing the Symbolic Math Toolbox, with a simple command—mupad. You type mupad on the command prompt and pops open a MuPAD Notebook. Now you are in the MuPAD fairy land. Naturally, the language changes, the syntax changes, the look and feel of everything changes. You can forget MATLAB here. To enjoy the new environs, you must give yourself enough time to learn the new tricks, new mannerisms, and new eccentricities. The exploration is rather deep.

For on-line help type:
`help mupad`

The MuPAD Notebook provides access to the full functionality of MuPAD and also serves as a complete record of your work that includes:

Your input commands that appear in red (see Fig. 8.1 and compare with Fig. 8.2).

MuPAD output appears in blue and follows the input command. Graphical output, including animations, gets embedded in the output too.

Text comments can be added before each command line and appear in black.

Thus, the notebook is a convenient way of not just doing computation in Mu-PAD but also preparing reports on your work simultaneously. You can save these notebooks and reopen them later to continue working from where you had left last time.

Note the following distinct features of the notebook layout (see Fig. 8.1):

- **Work area:** This is the main part of the Notebook window where you type commands and see the output.

- **Command Bar:** Located on the right of the work area, this bar contains a list of mathematical icons for most frequently performed calculations. Clicking on an icon types out the corresponding command in the work area with placeholders for the required input variables from the user.

- **The menu bar:** The menu bar on top of the window (below the name bar of the notebook) has the usual icons that you expect in all window interfaces. However, this bar is context sensitive and changes with what kind of object your cursor is placed on. When the cursor is on a graphical object that has animation, the menu bar brings up controls for playing the animation.

- **Command sensitive on-line help:** If you place your cursor on a command (i.e., name of a function) you have typed and left click the mouse, you get a menu from where you can access on-line help on that command, displayed in a separate MuPAD help window. You can always access the help window from the **Help** menu of MuPAD.

Three basic things about syntax

1. **Variable assignment** is done with ':=' and NOT with '=' as in MATLAB. Thus, you must type `f := x^2 +1` or `x := 5` in MuPAD notebook.

2. **Output suppression** is done with ':' at the end of the command and NOT ';' as in MATLAB. Thus, to suppress the output, type `f := (x^2 +1): `.

3. **Recalling the previous answer** is done with '%' and NOT ans as in MAT-LAB. Thus typing, `subs(%, x=a)` will substitute a for x in the previous answer.

8.5.1 Graphics and animation

MuPAD has extensive graphics capability. The number of graphics functions and the ease with which one can use various options for rendering graphs is outstanding. Strangely enough, most of these functions are not available through the Symbolic Math Toolbox. You must access these functions through the MuPAD Notebook. MuPAD provides a whole suite of graphics primitives, such as point, line, arc, box, circle, ellipse (both in 2-D and 3-D), and surface, cone, cylinder, sphere, etc.

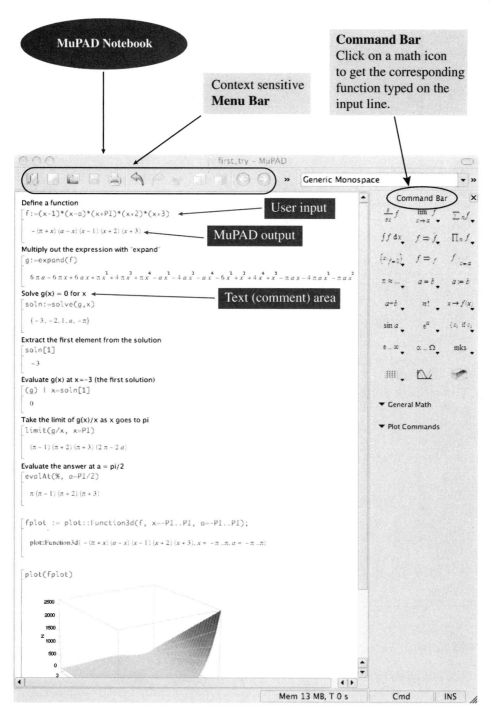

Figure 8.1: A typical session in a MuPAD Notebook. MuPAD notebooks are invoked by typing **mupad** on the MATLAB command line. Once the Notebook opens, you are in the native MuPAD environment. All commands typed here must confirm to MuPAD syntax. Compare these commands with those shown in Fig. 8.2 used in the Symbolic Math Toolbox for the same set of calculations.

```
>> syms x a pi
>> f = (x-1)*(x-a)*(x+pi)*(x+2)*(x+3)        Define a function f.
f =
-(pi + x)*(a - x)*(x - 1)*(x + 2)*(x + 3)

>> g = expand(f)                             Expand f and call it g.
g =
6*pi*a - 6*pi*x + 6*a*x + pi*x^2 + 4*pi*x^3 + ...

>> soln = solve(g,x)                         Solve g = 0 for x.
soln =

   -3

   -2

    1

    a

  -pi

>> soln(1)

ans =

-3

>> simplify(subs(g,x, soln(1)))             Substitute one of the
                                            solutions in g and see
ans =                                       if it satisfies g = 0.

0

>> limit(g/x,x,pi)                          Find  lim  g(x)/x .
                                                 x→π
ans =

(pi - 1)*(pi + 2)*(pi + 3)*(2*pi - 2*a)

>> simplify(subs(ans,a,pi/2))              Evaluate ans at a = π.

ans =

pi*(pi - 1)*(pi + 2)*(pi + 3)

>> ezsurf(f)    Make a surface plot of f.
```

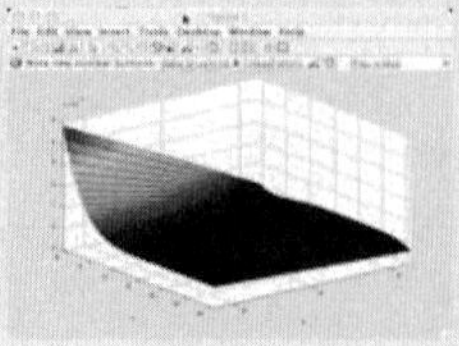

Figure 8.2: This simple set of calculations is carried out using the Symbolic Math Toolbox. Compare these commands with those shown in Fig. 8.1 carried out using the MuPAD Notebook.

With these primitives and their host of attributes, you can build very sophisticated graphics fairly quickly. The on-line help documentation (accessible through the MuPAD Notebook menu) is very good as it provides examples on each primitive.

Here, we mention the two basic workhorses of MuPAD graphics, `plotfunc2d` and `plotfunc3d`. Note that these functions are to be used in MuPAD Notebook. The only graphics functions available through the Symbolic Math Toolbox are those belonging to the *EZ* family—`ezplot`, `ezsurf`, `ezcontour`, etc. The function `plotfunc2d` can be used to plot any scalar function or functions, just like `ezplot`, but with a lot more control on the plot attributes (e.g., color, axes, ticks). First, we use the simplest form:

$$\texttt{plotfunc2d(sin(x\textasciicircum 2), x = 0..4*PI)}$$

to plot $\sin x^2$ over $x \in [0, 4\pi]$. The first argument is the function to be plotted (here, $\sin x^2$) and the second argument is the range of x to be used for plotting. Note that π is `PI` in MuPAD and not `pi` as in MATLAB. The output is shown in Fig. 8.3(a). The 3-D counterpart, `plotfunc3d`, is equally easy to use:

$$\texttt{plotfunc3d(sin(x\textasciicircum 2)+sin(y\textasciicircum 2), x = -PI/2..PI/2, y = -PI/2..PI/2).}$$

The output is shown in Fig. 8.3(b). So, what is the difference between these functions and `ezplot` functions? These functions can take a lot of optional arguments, literally tens of them for each object in the plot, e.g., for axes, line, text, labels. MuPAD on-line help provides enough examples to show you how to use these optional arguments. To activate the on-line help, click the left button of the mouse when the cursor is on the function `plotfunc2d` in the notebook, and choose **Help about 'plotfunc2d'**.

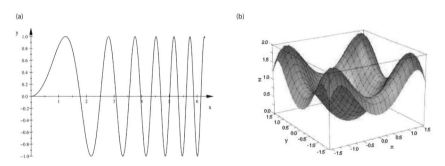

Figure 8.3: Plots produced by basic plotting commands (a) `plotfunc2d` and (b) `plotfunc3d`

The most interesting thing about these functions is how easily they let you animate a graph. All you need is an extra parameter in the mathematical function you are plotting. Provide a range for the parameter and watch the graph change dynamically as the parameter changes! Animating a graph, 2-D or 3-D, is as simple as that.

Let us see a simple example first. Let us plot $\sin(ax^2)$ as 'a' changes from small values to 1:

```
plotfunc2d(sin(a*x^2), x = 0..4*PI, a = 0.1..1)
```

Note the third argument—the range for the parameter a. That's it. No fanfare, no elaborate functional calls. MuPAD is smart enough to interpret that the range given for a is the animation parameter. Figure 8.4 shows three frames selected from the animation thus produced. These frames were taken by pausing the animation at a frame and exporting that frame as a graphics file (just left-click and select your options for exporting the graphics).

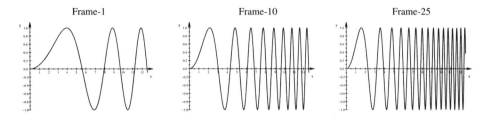

Figure 8.4: Three frames taken from the animation sequence produced by plotfunc2d command.

Needless to say that plotfunc3d will do the same thing with 3-D plots, given an extra parameter and its range. Whatever you see here, however, is just the tip of an iceberg. MuPAD gives you a lot more facility for animating graphs. The more you want, the deeper you have got to dig.

Both plotfunc2d and plotfunc3d are actually high-level plotting functions in MuPAD. Sophisticated animation requires working with graphics primitives. Mu-PAD provides an amazing suite of graphics primitives. The function plotfunc2d is itself made of the primitive 2-D plotting function plot::Function2d. You can use this primitive function to create many graphics objects without displaying them on the screen. Then you can use the plot function to display all these objects, either separately or together. As an example, let us say that we have two functions $F_{\text{mech}} = k * x$ and $F_{\text{elec}} = \frac{V^2}{10(1-x)^2}$, and we are interested in seeing the intersection of these two curves as V varies over a range of values. Just for the kicks, we will hatch the area between these two curves and see it change with the animation of the graphs. Here are the commands:

```
Fm := plot::Function2d(15*x, x=0..1, Color=RGB::Black):
Fe := plot::Function2d(0.1*V^2/(1-x)^2, x=0..1, V=2..5):
Diff := plot::Hatch(Fm, Fe):
plot(Fm, Fe, Diff, ViewingBoxYRange=0..15, Frames=20, TimeRange=0..3)
```

The last command is what produces the animation (by displaying the three graphics objects created by the first three lines). The optional arguments control the y-axis, the number of frames in the animation, and the time range over which the animation is played. Try it out. Three static frames taken from this animation are shown in Fig. 8.5.

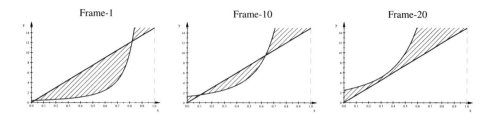

Figure 8.5: Three frames taken from the animation sequence produced by using three graphic objects produced with the primitives plot::Function2d and plot::Hatch, and displayed using plot.

8.6   Summary: Some Symbolic Math Toolbox Commands

Here is a list of the commands we have discussed. The things inside the parentheses are there as examples. You could change them.

`syms x t y a b`	Declare x, t, y, a, and b as symbolic. Do this first.
`sym([1 2 3])`	Treat the list [1 2 3] as symbolic.
`diff(sin(t),t)`	Calculate $\frac{d}{dt}\sin t$ to be $\cos t$.
`int(x^2, x)`	Calculate $\int x^2 \, dx$ to be $x^3/3$.
`limit(f, x, a)`	Find $\lim_{x \to a} f(x)$.
`expand((a+b)^3)`	Expand $(a+b)^3$ to $a^3 + 3a^2b + 3ab^2 + b^3$.
`collect(x^2 + 2*x^2)`	Collect $x^2 + 2x^2$ into $3x^2$.
`solve(x^3-y, x)`	Find the three cube roots of y.
`subs(x*y^2, y, a)`	Substitute $y = a$ into xy^2 to get xa^2.
`pretty(x^2)`	Print out x^2 on two or more easy-to-read lines.
`simplify(sin(t)^2+cos(t)^2)`	Use trig to simplify $\sin^2 t + \cos^2 t$ to 1.
`eval('sin(7)')`	Treat 'sin(7)' as a MATLAB command.
`double(sym(rand(3)))`	Get numbers from a symbolic array.
`vpa(f, n)`	Evaluate f using 'n' digit variable precision arithmetic.
`vectorize('x^2')`	Put dots in an expression (turn x^2 into x.^2).
`char(sym(t^2))`	Turn a symbolic expression into a char string.
`f=inline('x^2-x')`	Make f so that, say, $f(7)$ calculates $7^2 - 7$.
`f=matlabFunction(exp)`	Make anonymous function f for exp.
`ezplot(exp(t))`	The simplest way to plot e^t.
`funtool`	Bring up a symbolic graphing calculator.

9. Honorable Mentions

There are mainly three other significant facilities in MATLAB, which we mention here, but leave it to the reader to explore as and when required.

9.1 Debugging Tools

MATLAB supports a built-in debugger, which consists of several commands such as `dbclear`, `dbcont`, `dbdown`, `dbquit`, `dbstack`, `dbstatus`, `dbstep`, `dbstop`, `dbtype`, `dbup`, etc. You can use these commands to help you debug your MATLAB programs. You can write these commands in your M-file that you want to debug, or you can invoke them interactively by clicking on them in the editor/debugger window. See the on-line help under `debug` for details of these commands.

9.2 External Interface: Mex-files

If you wish to dynamically link your Fortran or C programs to MATLAB functions so that they can communicate and exchange data, you need to learn about Mex-files. Consult *MATLAB External Interfaces* [4] to learn about these files. The process of developing Mex-files is fairly complicated and highly system-dependent. You should perhaps first consider nondynamic linking with your external programs through standard ASCII data files.

9.3 Graphical User Interface

It is also possible to design your own graphical user interface (GUI) with menus, buttons, and slider controls in MATLAB. This facility is very useful if you are developing an application package to be used by others. You can build many visual "user-friendly" features in your application. For more information, consult *Creating Graphical User Interfaces* [5].

A. The *MATLAB* *Language* *Reference*

A.1 Punctuation Marks and Other Symbols

For on-line help type: `help punct`

, **Comma:** A comma is used to
  - separate variables in the input and output list of a function
    Example: `[t,x]=ode23('pend',t0,tf,x0)`
  - separate the row and column indices in a matrix
    Example: `A(m,n)`, `A(1:10,3)`, etc.
  - separate different commands on the same line
    Example: `plot(x,y)`, `grid`, `xlabel('x')`, etc.

; **Semicolon:** A semicolon is used to
  - suppress the MATLAB output of a command
    Example: `x=1:10; y=A*x;` etc.
  - separate rows in the input list of a matrix
    Example: `A=[1 2; 4 9]`

: **Colon:** A colon is used to specify range
  - in creating vectors
    Example: `x=1:10; y=1:2:100;` etc.
  - for matrix and vector indices
    Example: see Section 3.1.3
  - in for loops
    Example: `for i=1:20, x=x+i; end`

' **Right Quote:** A single right quote is used to transpose a vector or a matrix.
    Example: `symA=(A'+A)/2`

' ' **Single Quotes:** A pair of single right quote characters is used
      to enclose a character string.
      Example: `xlabel('time')`, `title('My plot')`, etc.

.   **Period:** Other than in numbers (for decimal point), a period is used
      in array operations.
      Example: `Asq = A.^2` (see page 73)

..   **Two Periods:** Two periods are used in `cd ..` command to access
      parent directory.

... **Ellipsis:** Ellipsis (three periods) at the end of a command denote
      continuation to the next line.
      Example: `x=[log(1:100) sin(v+a.*b) 22.3 23.0 ...`
                   `34.0 33.0 40:50 80];`

!   **Exclamation:** An exclamation preceding a *command* is used to send
      the local operating system command *command* to the system.
      This command is not applicable to Macs.
      Example: `!emacs newfile.m` invokes the local *emacs* editor.

@   **At:** @ character denotes a function handle used to
      - pass functions in the input list of other functions
        Example: `fzero(@myfunction,3)`
      - create anonymous functions
        Example: `f=@(t) t*sin(omega*t)`

%   **Percent:** A percent character is used to
      - mark the beginning of a comment, except when used
        in character strings
        Example: `% This is a comment`, but `rate = '8.5%'` is a string
      - denote formats in standard I/O functions `sprintf` and `fprintf`
        Example: `sprintf('R = %6.4f', r)`

%%   **Double Percent:** Two consecutive % characters are used to mark the
      beginning of a cell in cell scripts (for debugging and publishing).
      Example: `%% This is the header of a cell script`

( ) **Parentheses:** Apart from its obvious use in arithmetic operations
      (such as `a=5/(2+x*(3-i));` etc.), parentheses are used to
      - enclose matrix and vector indices
        Example: `A(1:5,2)=5;` `v=x(1:n-5);` etc.
      - enclose the list of input variables of a function
        Example: `[t,x]=ode23('pend', t0, tf, x0)`

{ } **Curly braces:** These braces are used to
      - enclose cell indices (see 127), e.g., `C{2,1} = 5`
      - enclose arguments in LaTeXcommands, e.g., `ylabel('x_{n+1}^{-2}')`.

[ ] **Square brackets:** Square brackets are used to
      - form and concatenate vectors and matrices
        Example: `v=[1 2 3:9];` `X=[v; log(v)];` etc.
      - enclose the list of output variables of a function
        Example: `[V,D]=eig(A);` etc.

A.2   General-purpose Commands

See Section 1.6.6

For on-line help type:
`help general`

Help and Query			
`lookfor`	Keyword search for help	`whatsnew`	Display ReadMe files
`help`	On-line help	`info`	Info about MATLAB
`helpdesk`	On-line HTML help	`ver`	MATLAB version info
`doc`	On-line HTML documentation	`syntax`	Help on command syntax
`demo`	Run demo program	`why`	Give philosophical advice

Command Window Control			
`clc`	Clear command window	`home`	Send cursor home
`format`	Set screen output format	`echo`	Echo commands in script file
`more`	Control paged screen output	`↑,↓`	Recall previous commands

Working with Files and Directories			
`pwd`	Show current directory	`delete`	Delete file
`cd`	Change current directory	`diary`	Save text of MATLAB session
`dir, ls`	List directory contents	`type`	Show contents of file
`mkdir`	Create a new directory	`!`	Access operating system
`rmdir`	Remove directory	`path`	List accessible directories

Variable and Workspace			
`clear`	Clear variables and functions	`length`	Length of a vector
`who,whos`	List current variables	`size`	Size of a matrix
`load`	Load variables from file	`pack`	Consolidate memory space
`save`	Save variables in Mat-file	`disp`	Display text or matrix

Start and Exit			
`matlabrc`	Master start-up file	`quit`	Quit MATLAB
`startup`	M-file executed at start-up	`exit`	Same as quit

Time and Date			
`clock`	Wall clock time	`etime`	Elapsed time function
`cputime`	Elapsed CPU time	`tic`	Start stopwatch timer
`date`	Date, month, year	`toc`	Read stopwatch timer

A.3   Special Variables and Constants

Constants		Variables	
pi	π (=3.14159...)	ans	Default output variable
inf	∞ (infinity)	computer	Computer type
NaN	Not-a-Number	nargin	Number of input arguments
i, j	Imaginary unit ($\sqrt{-1}$)	nargout	Number of output arguments
eps	Machine precision		
realmax	Largest real number		
realmin	Smallest real number		

A.4   Language Constructs and Debugging

For on-line help type:
`help lang`

See Section 4.3.

Declarations/Definitions		
script	function	global
nargchk	persistent	mlock

Interactive Input Functions		
input	keyboard	uimenu
ginput	pause	uicontrol

Control Flow Functions		
for	while	end
if	elseif	else
switch	case	otherwise
error	break	return

Debugging				
dbclear	dbcont	dbstep	dbstack	dbstatus
dbup	dbdown	dbtype	dbstop	dbquit

A.5   File Input/Output

For on-line help type:
`help iofun`

See Section 4.3.7.

File Opening, Closing, and Positioning						
open	fopen	fclose	fseek	ftell	frewind	ferror

File Reading and Writing						
fread	fwrite	fprintf	fscanf	fgetl	fgets	textread
xmlread	xmlwrite	wklread	wklwrite	saveas	print	publish

A.6   Operators and Logical Functions

See Section 3.2.

For on-line help type:
`help ops`

Arithmetic Operators			
Matrix Operators		*Array Operators*	
+	Addition	+	Addition
–	Subtraction	–	Subtraction
*	Multiplication	.*	Array multiplication
^	Exponentiation	.^	Array exponentiation
/	Left division	./	Array left division
\	Right division	.\	Array right division

Relational Operators		Logical Operators	
<	Less than	&	Logical AND
<=	Less than or equal	|	Logical OR
>	Greater than	~	Logical NOT
>=	Greater than or equal	xor	Logical EXCLUSIVE OR
==	Equal		
~=	Not equal		

Logical Functions			
all	any	exist	find
finite	isempty	isinf	isnan
ismember	issparse	isstr	isfinite

A.7   Frequently Used Math Functions

For on-line help
type:
`help elfun`

See Section 3.2.4 for description and examples.

Trigonometric Functions			
sin, sind	asin, asind	sinh	asinh
cos, cosd	acos, acosd	cosh	acosh
tan, tand	atan, atand	tanh	atanh
	atan2		
cot, cotd	acot, acotd	coth	acoth
sec, secd	asec, asecd	sech	asech
csc, cscd	acsc, acscd	csch	acsch

Exponential Functions			
exp	log	log10	sqrt
expm1	log1p	log2	nthroot
pow2	nextpow2	realpow	reallog

For on-line help
type:
`help specfun`

Complex Functions			
abs	angle	conj	complex
real	imag	unwrap	cplxpair

Round-off Functions			
fix	floor	ceil	round
rem	sign	mod	

Specialized Math Functions			
bessel	bessely	besselh	beta
betain	betaln	ellipj	ellipke
erf	erfinv	gamma	gammainc
legendre	rat	dot	cross

A.8   Matrices: Creation and Manipulation

See Section 3.1.

For on-line help type:
`help elmat`

Elementary Matrices			
eye	ones	zeros	rand
randn	linspace	logspace	meshgrid

Specialized Matrices			
compan	hadamard	hankel	hilb
invhilb	magic	pascal	rosser
toeplitz	vander	wilkinson	gallery

Matrix Manipulation Functions			
diag	fliplr	flipud	reshape
rot90	tril	triu	:

For on-line help type:
`help matfun`

Matrix (Math) Functions			
expm	logm	sqrtm	funm

Matrix Analysis			
cond	det	norm	null
orth	rank	rref	trace
eig	balance	poly	hess

Matrix Factorization and Inversion			
chol	cholinc	lu	luinc
eig	eigs	svd	svds
qr	qz	schur	pinv

Sparse Matrix Functions: There are also several functions for creating, manipulating, and visualizing sparse matrices. Some of these are `spdiag`, `speye`, `sprandn`, `full`, `sparse`, `spconvert`, `spalloc`, `spfun`, `condest`, `normest`, `sprank`, `gplot`, and `spy`. See on-line help for complete listing.

For on-line help type:
`help sparfun`

A.9   Character String Functions

For on-line help
type:
`help strfun`

See Section 3.3.

General String Functions					
abs	char	eval	setstr	strcat	strvcat
string	strcmp	lower	upper	isstr	ischar

String $\Longleftrightarrow$ Number Conversion				
int2str	num2str	sprintf	dec2hex	mat2str
str2num	sscanf	hex2dec	hex2num	dec2bin

A.10   Graphics Functions

For on-line help
type:
`help graphics`
`help graph2d`
`help graph3d`

See Chapter 6.

EZ Graphics				
ezplot	ezpolar	ezcontour	ezcontourf	ezgraph3
ezplot3	ezmesh	ezmeshc	ezsurf	ezsurfc

2-D Graphics				
plot	loglog	semilogx	semilogy	fplot
bar	errorbar	compass	feather	stairs
polar	fill	hist	rose	quiver

3-D Graphics				
plot3	fill3	mesh	meshc	meshz
surf	surfc	surfl	cylinder	sphere
Contour Plots				
contour	contour3	contourc	clabel	pcolor
Volumetric Plots				
slice	isosurface	isocaps	isocolors	cotourslice
coneplot	streamline	streamtube	streamlice	streamparticles

Graphics Annotation				
xlabel	ylabel	zlabel	title	legend
text	gtext	grid	plotedit	rectangle

Axis Control and Graph Appearance				
axis	colormap	hidden	shading	view

Window Creation and Control				
clf	close	figure	gcf	subplot

Axis Creation and Control				
axes	axis	caxis	cla	gca

Handle Graphics Objects and Operations				
axes	line	patch	surface	text
figure	image	uicontrol	uimenu	
delete	drawnow	get	reset	set

Animation and Movies				
comet	getframe	movie	moviein	avifile
movie2avi	frame2im	im2frame	rotate	rotate3D

Hard Copy and Miscellaneous				
print	orient	printopt	ginput	hold

Color Control and Lighting				
caxis	colormap	flag	hsv2rgb	rgb2hsv
bone	copper	gray	hsv	pink
cool	hot	shading	brighten	diffuse
surfl	specular	rgbplot		

For on-line help type:
`help color`

A.11   Some Applications Functions

For on-line help
type:
help datafun

A.11.1   Data analysis and Fourier transforms

Basic Statistics Commands				
mean	median	std	min	max
prod	cumprod	sum	cumsum	sort

Correlation and Finite Difference				
corrcoef	cov	del2	diff	gradient

Fourier Transforms				
fft	fft2	fftshift	ifft	ifft2
abs	angle	cplxpair	nextpow2	unwrap

Filtering and Convolution				
conv	conv2	dconv	filter	filter2

For on-line help
type:
help polyfun

A.11.2   Polynomials and data interpolation

Polynomials				
poly	polyder	polyfit	polyval	polyvalm
conv	deconv	residue	roots	

Data Interpolation				
interp1	interp2	interpft	interpn	griddata

Fourier Transforms				
fft	fft2	fftshift	ifft	ifft2
abs	angle	cplxpair	nextpow2	unwrap

Filtering and Convolution				
conv	conv2	dconv	filter	filter2

For on-line help
type:
help funfun

A.11.3   Nonlinear numerical methods

Functions					
fmin	fmins	fminbnd	fminsearch	fzero	trapz
quad	quadl	dblquad	bvp4c	pdepe	dde23
ode23	ode45	ode113	ode23t	ode23s	odefile

Bibliography

[1] *MATLAB® 7 Desktop Tools and Development Environment*, The MathWorks, Inc., 2004.

[2] *MATLAB® 7 Mathematics*, The MathWorks, Inc., 2009.

[3] *MATLAB® 7 Programming Fundamentals*, The MathWorks, Inc., 2009.

[4] *MATLAB® External Interfaces*, The MathWorks, Inc., 2009.

[5] *Creating Graphical User Interfaces*, The MathWorks, Inc., 2009.

[6] *Using MATLAB® 7 Graphics*, The MathWorks, Inc., 2009.

[7] *MATLAB® 7 Release Notes*, The MathWorks, Inc., 2009.

[8] Kernighan, B. W., and D. M. Ritchie, *The C Programming Language*, second edition, Prentice Hall Inc., 1993.

[9] Strang, G., *Linear Algebra and Its Applications*, third edition, Saunders HBJ College Publishers, 1988.

[10] Golub, G. H., and C. F. Van Loan, *Matrix Computations*, third edition, The Johns Hopkins University Press, 1997.

[11] Horn, R. A., and C. R. Johnson, *Matrix Analysis*, Cambridge University Press, 1985.

[12] Gerald, C. F., and P. O. Wheatley, *Applied Numerical Analysis*, fifth edition, Addison Wesley Publishing Company, 1994.

[13] Press W., B. Flannery, S. Teudolsky, and W. Vetterling, *Numerical Recipes in C: The Art of Scientific Computing*, second edition, Cambridge University Press, 1992.

Index